Distributed Sensor Arrays
Localization

Distributed Sensor Arrays
Localization

By
Prabhakar S. Naidu

CRC Press is an imprint of the
Taylor & Francis Group, an **informa** business

CRC Press
Taylor & Francis Group
6000 Broken Sound Parkway NW, Suite 300
Boca Raton, FL 33487-2742

First issued in paperback 2022

ISBN-13: 978-1-138-63213-4 (hbk)
ISBN-13: 978-1-03-233948-1 (pbk)
DOI: 10.1201/b22108

Publisher's Note

The publisher has gone to great lengths to ensure the quality of this reprint but points out that some imperfections in the original copies may be apparent.

Visit the Taylor & Francis Web site at
http://www.taylorandfrancis.com

and the CRC Press Web site at
http://www.crcpress.com

Dedication

This work is dedicated to the memory of my mentor, the late Professor B.S. Ramakrishna, who shaped my career in the Department of Communication Engineering, Indian Institute of Science, Bangalore, India.

Contents

Preface

Uniformly distributed sensor arrays are impractical, as there is every likelihood that placement errors will lead to the response function veering off the theoretical response. In the extreme case, the sensors may be arbitrarily distributed, as when an array is launched from the air. Such an array is a distributed sensor array (DSA), the subject matter of the present work. The primary objective of any array processing is simply the localization of one or more transmitters and the estimation of the transmitted signal. In DSA, there is an additional twist as the sensor position is unknown. We need to localize both transmitter and sensors from the same observations. A DSA may consist of a few tens to a few hundred sensors. A sensor may be able to communicate with its immediate neighbor(s) and relay the information it has received from a transmitter(s).

Different types of information of interest from the point of localization are covered in Chapter 2. The most important and relatively simple ones are relative signal strength (RSS), time of arrival (ToA), and time difference of arrival (TDoA), which we discuss in Chapters 3 and 4. When a transmitter is moving with respect to an array (generally assumed to be stationary), there is an additional piece of information known as frequency difference of arrival (FDoA). A transmitter is the point source and the wavefield is sampled by a randomly spaced sensor array. This situation is an ideal candidate for the application of compressive sensing, which we describe in Chapter 4. Two or more parameters may be estimated at the same time using the sensor outputs, either directly or recursively, as in an expectation–maximization (EM) algorithm. This aspect is covered in Chapter 5. Finally, in Chapter 6, we look into the possibility of self-localization with topics, such as cooperative localization and multidimensional scaling. We also look into clock synchronization, in particular how to reduce the effect of synchronization error by means of single- or double-reply approaches.

This work will be of interest to those working in the area of array signal processing, particularly when confronted with large distributed arrays. As the book covers practically all popular and not-so-popular methods, it will be of interest to all those who are just approaching the topic of distributed sensor array. This volume can also be used as a reference book for courses on array signal processing.

I have maintained the same style of presentation as in my previous works. All ideas are presented in mathematical language with illustrative examples, wherever required.

Acknowledgments

The significance of distributed sensor arrays struck my mind while teaching and researching regular arrays and their applications for over a decade. But in the real world, arrays are rarely perfect. Small to large imperfections, very common in practical setup, can lead to serious errors in the modeled response. This has led to sensor array where all sensors are indeed assumed as randomly distributed. The problem now turns out to be localization of both transmitter and sensors. I pursue this line of research at MVGR College of Engineering in Vizianagram, Andhra Pradesh, India, as a visiting professor. Many students contributed to this endeavor. It is difficult to name them all, but here are a few names of those who made important contributions: Soujanya, Kiran, Keertan, Mrudula, Surekha, Tejaswi, Vysalini, Venkataramana, Sagarika, Anuradha, Prithviraj, Sagar, and Chaitenya. Finally, I would like to thank the management of MVGR College, in particular, Professor K.V.L. Raju, for giving me this opportunity.

Author

Prabhakar Naidu has pursued research and teaching in signal processing for over 30 years at the Indian Institute of Science, Bangalore, India. His basic training has been in geophysical signal processing. His earlier published books are *Modern Spectral Analysis of Time Series* (CRC Press, 1996), *Analysis of Geophysical Potential Fields: A DSP Approach* (Elsevier, 1998), *Modern Digital Signal Processing: An Introduction* (Narosa, 2003), and *Sensor Array Signal Processing* (CRC Press, 2001). Dr. Naidu is currently a visiting professor at MVGR College of Engineering, Vizianagram, Andhra Pradesh, India, and his e-mail address is psnaidu101@gmail.com.

1 Distributed Sensor Arrays

There are many engineering applications where it is difficult to build a regular array because of practical limitations, in the case of a forest fire or for flood control, for example. During war, to monitor enemy troop movement, wireless sensors are often used by projecting them into enemy territory. In such a situation, it is impossible to build a regular sensor array. Likewise, in wireless communications, non-invasive localization of a tumor in biomedical investigation, when tracking a moving vehicle, leakage of crude oil from a pipeline, or in seismic exploration for hydro-carbons, placement of sensors is often constrained by the accessible space.

This chapter explores different localization techniques where the sensors are randomly distributed. Since time synchronization is a difficult task in distributed sensor arrays (DSAs), we emphasize methods that do not require time synchronization. Toward this end, we provide details of localization methods based on received signal strength, lighthouse effect, frequency difference of arrival (FDoA) measurements, and phase difference estimation.

DSAs are considered wired when all sensors are connected to a central processor, either through physical wire or microwave communication (wireless). In the first category, we have a regularly shaped array, for example, linear, rectangular, or circular arrays of known geometry. There are irregularly distributed but wired sensors; for example, randomly distributed sensors at known locations, connected to a central processor. When the array is very large and is populated with many tiny sensors with limited power source, it is no longer possible for sensors to individually communicate to the central processor. Either a sensor communicates with its nearest neighbor or at utmost with a local processor, also known as an anchor. But the sensor position is generally not known. Such a sensor array is known as an ad hoc array. In this chapter, we shall consider the response functions of various types of sensor distributions.

1.1 REGULAR SENSOR ARRAY

A standard sensor array is a linear, rectangular, or circular array with equispaced sensors. There are variations of these geometries devised for specific applications, such as an L-shaped array in seismic monitoring or a towed array for naval applications.

Some examples of regular planar array of sensors are shown in Figure 1.1. An important requirement is that the sensor spacing and the geometry of the array must be precisely controlled. Phase difference between successive sensors must be constant and its magnitude must be less than π. This requires the sensor spacing to be less than half-wavelength, $\lambda/2$ (except for circular array [1]). Random errors in the sensor spacing can result in large side lobes in the array response. Similarly, individual sensor response, required for defining the steering vector, must be carefully estimated. This is known as array calibration. These are some of the nagging problems faced by any sensor array designer. These issues have been discussed in an earlier companion publication [2]. In the present work, we relax some of these

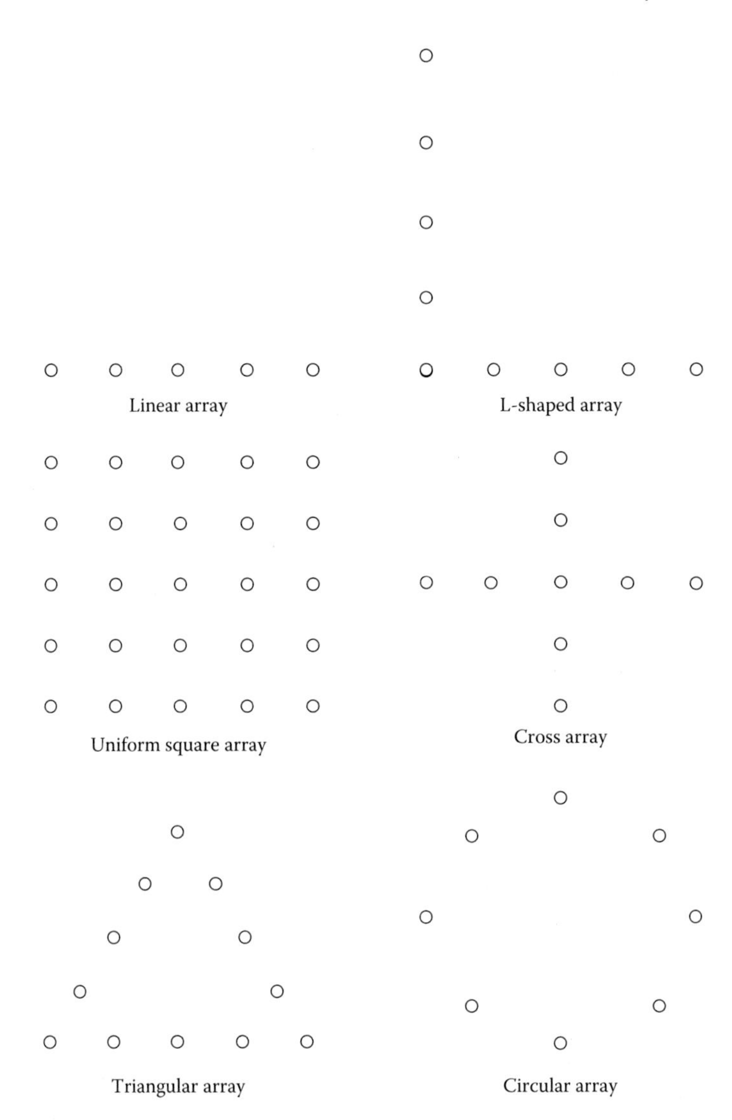

FIGURE 1.1 Some examples of regular planar arrays.

stringent requirements. The sensors are no longer required to be arranged in any specific order. Often, we have many redundant sensors, allowing a kind of trade-off between a precisely controlled array of a small number of sensors versus an array of a large number of randomly placed sensors.

1.2 WIRED SENSOR ARRAY

In many applications, it is not possible to achieve an ordered distribution of sensors, particularly when we have a very large number of sensors over a large area. But the location of sensors is precisely known. A recent example of such a large array is the largest radio telescope in the world, consisting of 25,000 small antennas distributed over Western Europe (aperture $\approx$ 1000 km). Sometimes, the constraints of space, for example in biomedical applications, demand sensors to be distributed over available space. In a distributed array of the previously mentioned type, all sensors are wired (or alternatively connected wirelessly) to a central processing unit where most of the signal processing tasks are carried out. Sensors, however, perform signal conditioning, A/D conversion, and so on. Each sensor is also able to communicate with the central processor. The location of all sensors is known, but the location of the transmitter and the emitted signal waveform are unknown. Wired sensor arrays of this type are usually small, with a limited number of sensors strategically placed surrounding a target of interest. Sometimes it becomes necessary to keep the target away from the sensor array for safety or security reasons. Location estimation is based on ToA information at each sensor. ToA estimation requires the transmitter to send out a known waveform at a fixed time instant. The time of arrival at each sensor is measured relative to the instant of transmission. The clocks of all sensors and the clock at transmitter must be synchronized to get correct ToA estimates. In addition, the transmitter must be a friendly one so that it may be programmed to transmit a waveform at a preprogrammed time instant. When this is not possible (e.g., with an enemy transmitter), we can use TDoA at all pairs of sensors whose clocks are synchronized. One advantage of using of ToA or TDoA for localization is that it is not necessary to maintain $\lambda/2$ spacing between sensors as in a regular array. Nor is it necessary to calibrate sensors.

When a sensor is not equipped to estimate ToA/TDoA, the received signal would be sent to the central processing unit, which will then take over the tasks of ToA/TDoA estimation and target localization. In this case, the ToA/TDoA will include the time of communication from the sensor to the central processing unit. As the distance between each sensor and the central processing unit (anchor) is known, it is easy to estimate this communication time and subtract it from the measured ToA. This correction, however, is insignificant when we are dealing with acoustic or seismic arrays—as the speed of a sound wave is much lower than that of electromagnetic waves—used for communication from a sensor to the central processing unit.

1.2.1 DSA Response

As in a regular sensor array, we can think of directivity or the beam pattern of a DSA. In a simple delay-and-sum beam formation, the outputs of all sensors, after suitable delay, are summed up to estimate the signal emanating from the chosen test point. Such a beam is also known as a focused beam [2]. The output power in the beam will represent a measure of our ability to detect the presence of a transmitter at the focal point. Thus, we can compute the array response at each point of space in the region of interest around the sensors. A large power peak at some point reveals the probable presence of a transmitter at that location. Indeed, this is the basis of mapping microwave energy emitted by numerous stars. Preparation of such an energy map is a highly computer-intensive task requiring a large mainframe-dedicated computer system. It is unlikely that a large computing power will be available in most applications where a DSA is employed. Moreover, there is the other serious issue of large side lobes found in the directivity pattern of random array. To demonstrate this, we shall consider the response of a randomly distributed sensor array (see Figure 1.2). The sensors cover a square region with a dimension of 4λ × 4λ where λ is the wavelength. The transmitter is placed at (40λ, 0). We have computed the response of the random array and a regular square array of the same size. The response (magnitude) was computed along the y-axis passing through the transmitter position.

The response functions are shown in Figure 1.3 for both randomly scattered and regularly distributed sensors. It is interesting to observe that, while the main lobes are identical, the side lobes of the scattered array are much larger.

1.2.2 Focused Beam

Each sensor output is delayed by an amount equal to the computed ToA. The sum of all delayed sensor outputs is calculated thus:

$$
\begin{aligned}
x_{foc}(t) &= \frac{1}{N}\sum_{i=1}^{N} x_i(t) = \frac{1}{N}\sum_{i=1}^{N}\int_{-\infty}^{\infty} X(\omega)e^{j\omega(t+\tau_i)-\omega\frac{r_i}{c}}d\omega \\
&= \int_{-\infty}^{\infty} X(\omega)e^{j\omega t}\frac{1}{N}\sum_{i=1}^{N} e^{j\omega(\tau_i-\frac{r_i}{c})}d\omega \\
&= \int_{-\infty}^{\infty} X(\omega)e^{j\omega t}H_{foc}(\omega)d\omega
\end{aligned}
$$

where $r_i(\mathbf{r})$ is the distance between the test location $\mathbf{r}$ and the ith sensor, c is the wave speed, τ_i is the observed ToA at the ith sensor, $X(\omega)$ is …, and

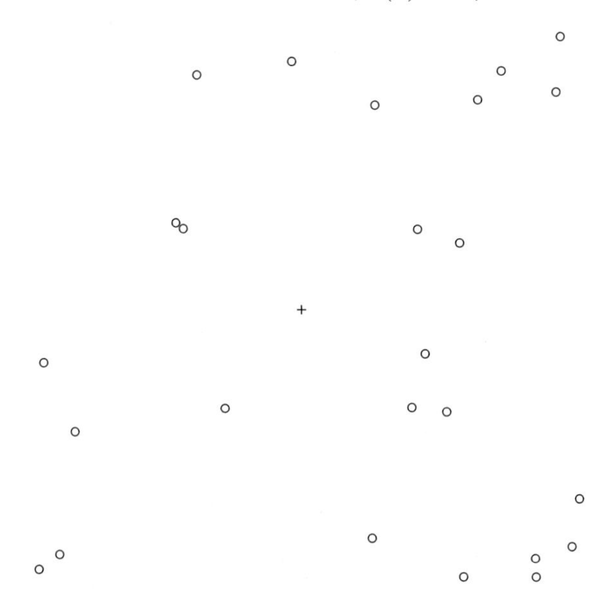

FIGURE 1.2 Randomly placed sensor array. Sensors (25 sensors) are scattered over a square space of the size 4λ × 4λ. Sensor locations are known.

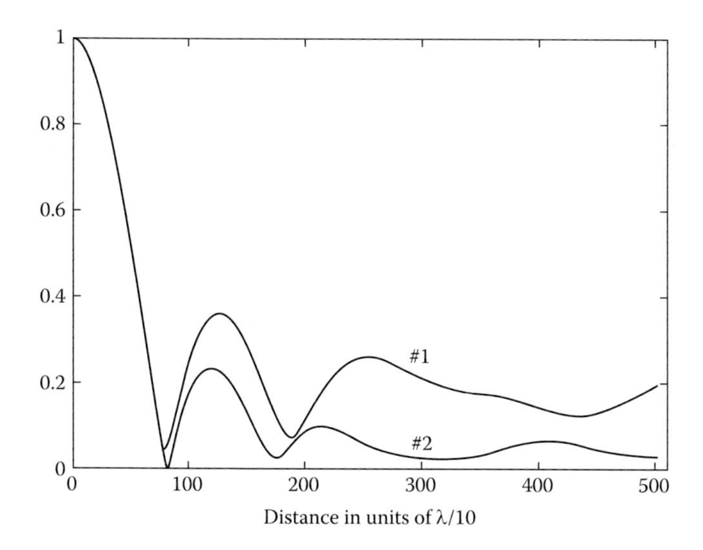

FIGURE 1.3 Responses of randomly distributed sensor array of 100 sensors (curve #1) and regular square array of 100 sensors in the same area (curve #2). The transmitter is placed at the far field point (40λ,0). Response is computed along the y-axis passing through the transmitter location. Notice the presence of large side lobes in the case of the randomly distributed sensor array.

$$H_{foc}(\omega) = \frac{1}{N}\sum_{i=1}^{N} e^{i2\pi(\omega_0 \tau_i - \omega_0 \frac{r_i}{c})}$$

This is the response function of the focusing action. The magnitude of the response function is of interest for localization of the transmitter. With $|H_{foc}(\omega)| \leq 1 \quad \forall\omega$ except at the actual location of the transmitter, where it is equal to one.

The response function is computed using the previously estimated ToA from the transmitter to every individual sensor. The ToA may be estimated by cross-correlating sensor output with the transmitter output, which is known. Let $\tau_i = 1,\ldots N$ be actual ToA at N sensors. Consider a transmitter at some test location $\mathbf{r}$, where we wish to compute the response. Compute the ToA $r_i(\mathbf{r})/c, i = 1,\ldots,N$, from the test location to each individual sensor. The magnitude–response functions of the random and uniform arrays are shown in Figure 1.3. While the main lobe responses are identical, the side lobe response of the random array is much higher. The side lobe structure, however, depends upon the actual placement of sensors. The side lobe structure also depends upon the number of sensors. In Figure 1.4, responses of two random arrays, with 100 sensors (curve #1) and 1000 sensors (curve #2), are shown. The size of arrays is the same as in Figure 1.1; that is, 4λ × 4λ. In Figure 1.5, we show the response when the transmitter is at the center of an array with 100 randomly distributed sensors in a 4λ × 4λ square.

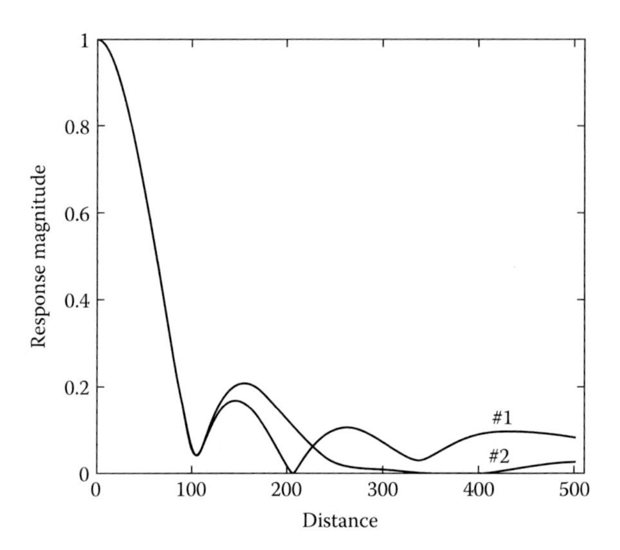

FIGURE 1.4 Response of randomly distributed array of 100 sensors (curve #1) compared with that of 1000 sensors (curve #2). Array dimensions and other parameters are the same as in Figure 1.3.

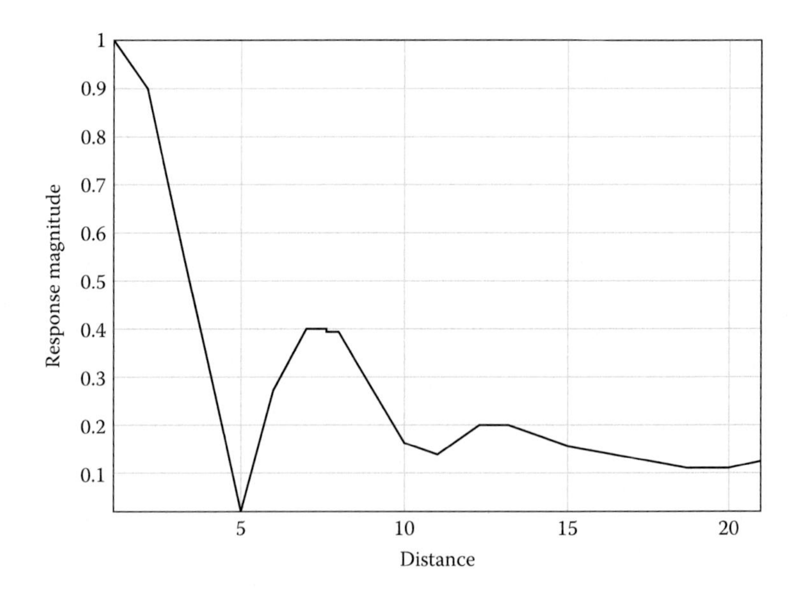

FIGURE 1.5 Response (magnitude) when the transmitter is at the center surrounded by 100 sensors uniformly distributed inside a square of $4\lambda \times 4\lambda$. Response is computed along the y-axis passing through the transmitter location. The distance is measured in units of $\lambda/10$.

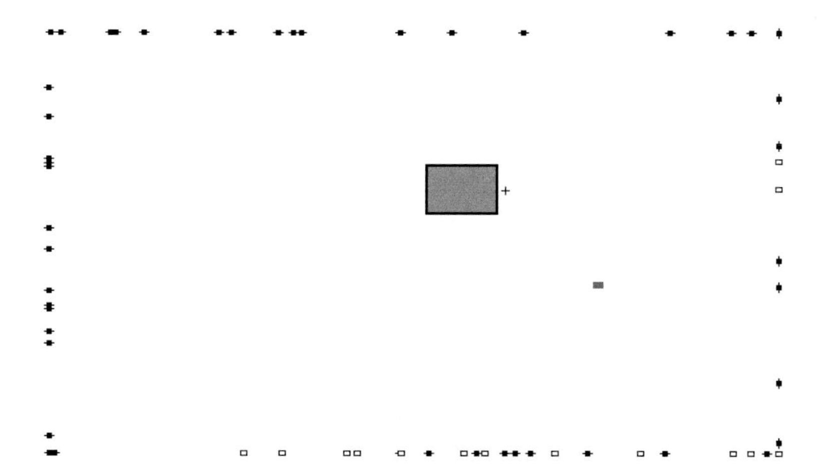

FIGURE 1.6 Rectangular boundary array of size $4\lambda \times 2\lambda$. There are 64 sensors on the boundary plus one reference sensor close to the control room (large gray square). All ToA are measured with respect to the reference sensor. The small gray block represents the transmitter (intruder).

1.2.3 Boundary Array

We consider sensors distributed over a boundary, for example, over a compound wall enclosing a secured building. The goal here is to detect and track an intruder in real time. A similar situation will arise in monitoring a reactor to avoid a possible explosion caused by excessive heat. Let the sensors be placed on the rectangular boundary but at arbitrary locations, as in Figure 1.6. An intruder, as he walks or whispers, generates an acoustic signal that is well within the reach of the acoustic sensors. The waveform received by each sensor is sent to a central processor. It is assumed that all sensors are wired to the central processor. As in distributed planar array, the processor essentially computes focused beams and estimates regions of energy concentration, which are likely locations of one or more intruders. The response of the array along the x-axis passing through the transmitter is shown in Figure 1.7, and that along the y-axis is shown in Figure 1.8. Distance, in both figures, is measured in units of $\lambda/25$. Notice that the responses along the x-axis and y-axis are quite different. There are many large side lobes and one sharp main lobe in the response function along the x-axis, but there is only a wide main lobe without any side lobes in the response function along the y-axis.

1.3 AUTONOMOUS WIRELESS SENSOR NETWORK

There is a growing need for low-powered sensors capable of sensing the surrounding environment and communicating information, often to the nearest neighbor. Many such sensors (a few hundreds to thousands) are used for monitoring inaccessible, hostile, or dangerous environments. Such wireless DSAs are often known as ad hoc sensor networks and sensors as nodes. Here "ad hoc" refers to the fact that the network is not preplanned. The sensor position is unknown as it may also be constantly changing to meet the changing risk assessment. Sensors are often dropped from the air or projected into

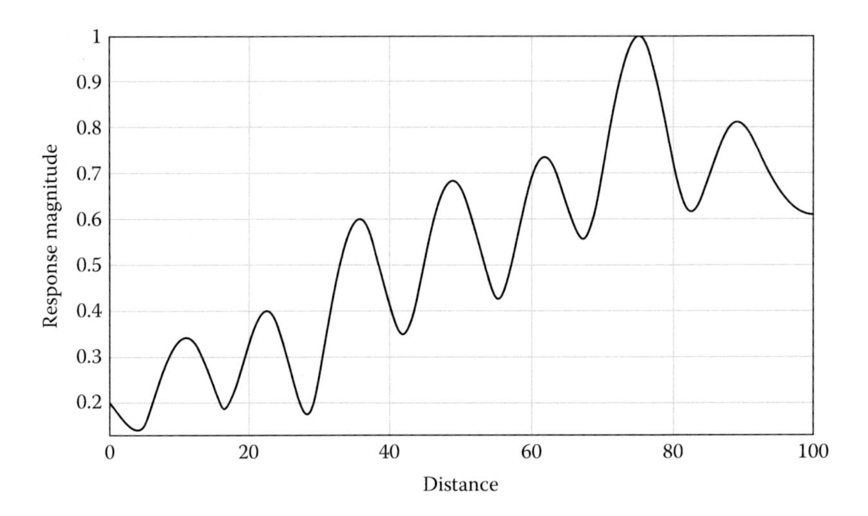

FIGURE 1.7 Response (magnitude) of boundary array along x-axis passing through the transmitter, which is located at (3λ, 0.8λ). Distance is measured in units of λ/25.

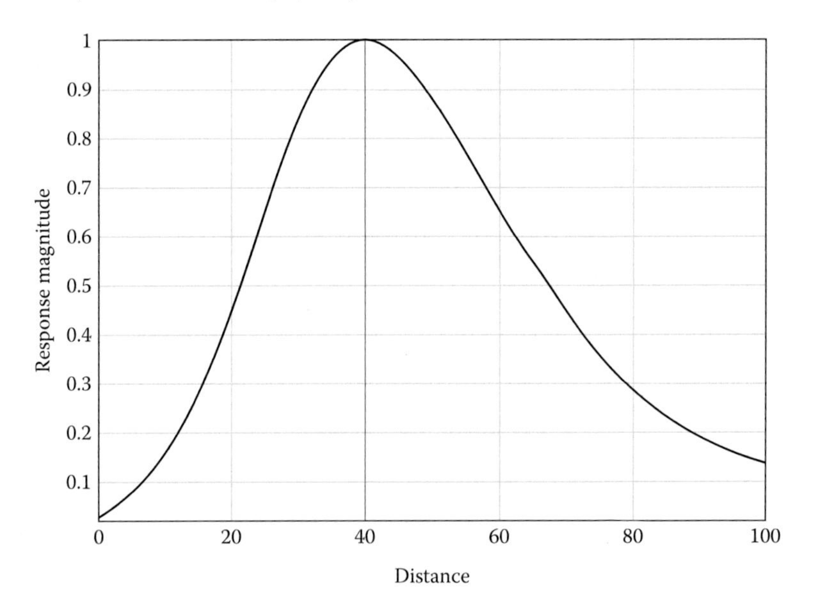

FIGURE 1.8 Response (magnitude) of boundary array along y-axis passing through the transmitter, which is located at (3λ, 0.8λ). Distance is measured in units of λ/25.

enemy territory using firepower. Each sensor acts on its own without being under the command control of a central computer. Thus, the sensors are autonomous. However, they are within communication distance of their immediate neighbors. All sensors in an ad hoc sensor network depend on their own battery power supply to receive and transmit capability. Hence, they have to be extremely frugal in power consumption. Each sensor carries one or more sensing elements. Micromachining has allowed researchers to pack many microsensors into a small space while maintaining performance levels similar to or better than those of conventional sensors. These include thermal sensors, accelerometers, microphones, radiation detectors, and so on. Because of power limitations and also cost, it is not possible to employ a global positioning system (GPS) receiver for position information, which is essential in any environment-probing mission. However, in any large sensor network, there are a few anchors or reference nodes whose location is precisely known, probably through a GPS receiver. Since the sensors (other than anchor sensors), also known as blind sensors, are necessarily inexpensive—for the obvious reason of cost reduction—they have only limited computing power. This means that the blind sensors can carry out only simple calculations. Naturally, all advanced signal-processing techniques for direction finding, ToA estimation, and so on, are out of reach for the blind sensors. All blind sensors have a radio frequency (RF) receiver and a transmitter (transceiver), but with limited power. They can only receive/transmit a message to/from their immediate neighbors. Each blind sensor has an environment-sensing (e.g., temperature) unit and a range estimator unit. An on-board processor with storage memory controls all three units. A schematic diagram of a typical blind sensor is shown in Figure 1.9. The transceiver shown in Figure 1.9 uses RF radiation.

Apart from the large power requirement, the size of the antenna needs to be a significant multiple of the wavelength to be at all efficient. To achieve milliradian collimation at 1 GHz, as in an inexpensive laser pointer, would require a 100 m antenna [4], which is indeed impossible in any tiny sensor used in a wireless sensor network (WSN). Another disadvantage of RF communication is rapid decay of power with range.

For ground-based communication, such as mobile telephones, the average power decay is $\propto 1/\text{distance}^4$.

Free-space optical communication provides an attractive alternative to wireless communication. Laser diodes and mirrors can be made very small, of the order $-0.03\ \mu m^3$. Another advantage of optical communication is low power decay, of the order $\propto 1/\text{distance}^2$. There are many other benefits accruing from the use of

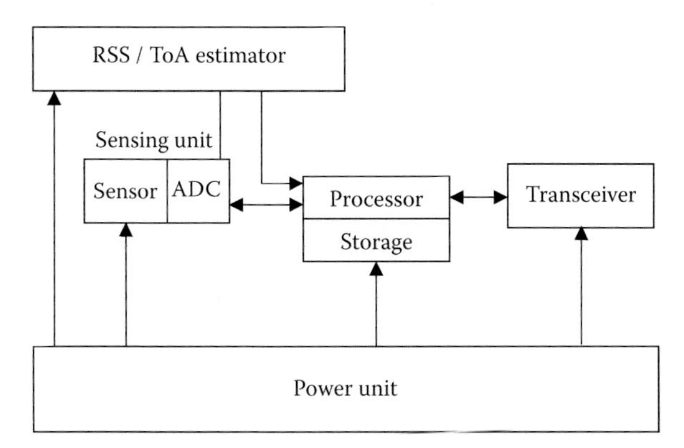

FIGURE 1.9 Basic structure of a sensor node. There are three units: (a) range estimator, (b) environment sensors, and (c) transceiver. All three units are connected to a processor with memory. A power unit consisting of a battery supplies power for all three units and the processor. (Adapted from L. M. P. Leao de Brito, and L. M. R. Peralta, *Polytechnical Studies Review*, vol. 8, no. 9, 2008, pp. 1–27 [3].)

optical communication. The only disadvantage is the need for line-of-sight communication. There are two approaches to free-space optical communication: the passive reflective system and the active-steered laser system.

The passive reflective system consists of three orthogonal mirrors that form the corner of a cube, hence the name corner-cube retro reflector (CCR), whose interesting property is that a beam entering from any direction is reflected back parallel to the incident beam (Figure 1.10). By tilting a single mirror of the CCR by angle δ, the direction of the outgoing beam will be offset. When the bottom mirror is actuated by 5.7 mrad, the return beam is split into two beams, each directed 11.4 mrad away from the optical axis, well away from the telescope entrance aperture [5]. A lens is used to collect the diffracted light and converted into electrical pulses (see Figure 1.11).

Since CCR is a passive device, it does not draw any power from the sensor. Power is derived from the radiator (i.e., anchor). This arrangement is not only quite suitable for the WSN, but it can also be used to estimate the range from received signal strength (RSS), which we shall discuss in Chapter 2. The laser rangefinder, if available in the anchor node, can be used to estimate the range to each sensor node. Location estimation will, however, require three or more anchor nodes communicating with the same CCR. Note that in this passive approach, node-to-node communication is not possible. However, in the active beam-steered method, there is no such limitation. A very low-power-consuming on-board light source, such as a vertical cavity surface emitting laser (VCSEL), a collimating lens, and MEMS beam-steering optics [4], are used to send a tightly collimated light beam toward an intended receiver, thus facilitating peer-to-peer communication [4].

1.3.1 Wireless Sensor Network (WSN)

The anchor sensors have the benefit of a large power battery; hence, they are equipped with a GPS receiver, on-board computer, radio frequency (RF) beacon, and so on. Each anchor broadcasts the signal to as many blind sensors as possible and enables them to estimate their relative distances from the transmitting anchor. When a blind sensor can measure its relative distance from three or more anchors, it is possible to estimate its position coordinates using a method of triangulation or multilateration. Localization (along with time synchronization) of sensors is essential in any application of an autonomous WSN. Some well-known applications are

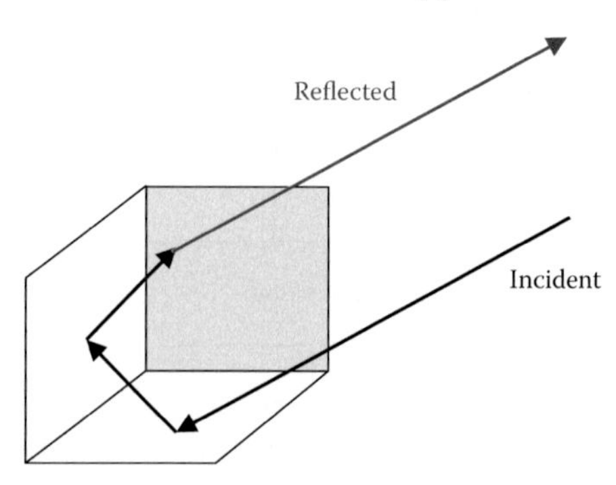

FIGURE 1.10 Corner-cube retro reflector. A beam entering from any direction is reflected back parallel to the incident beam.

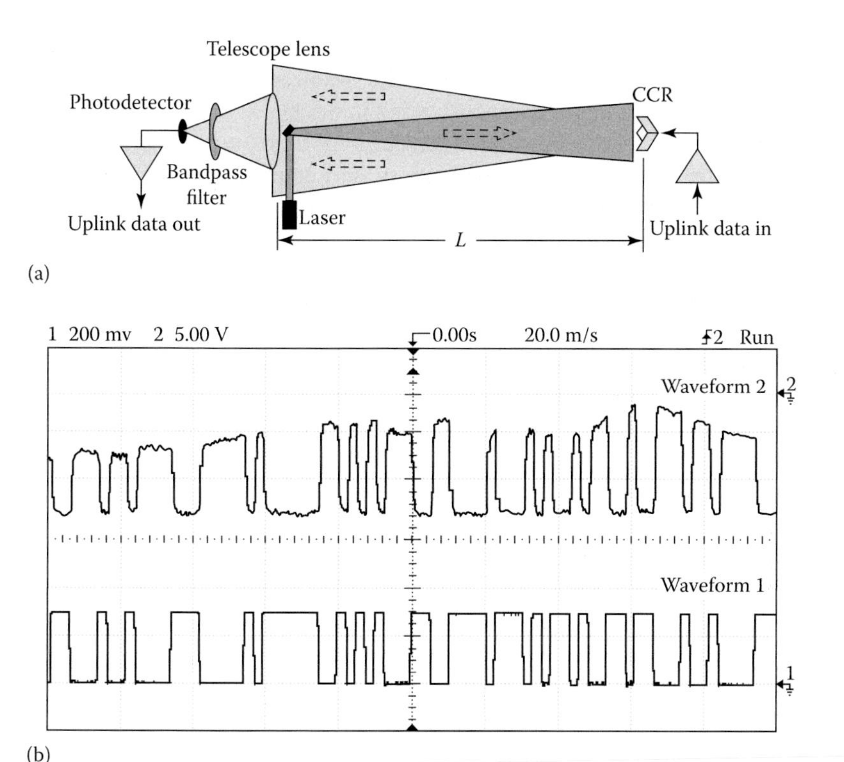

FIGURE 1.11 Corner-cube reflector installed in the sensor is used to reflect the laser beam back to the transmitter after modulating it with data bits from the sensor; (a) Optical setup with CCR for uplink free-space optical communication and (b) Results of transmission. Waveform 1 is transmitted bits. Bottom mirror is modulated with this sequence. Waveform 2 is detected 400 bits/sec at a distance of 180 m. (From Zhou, L. et al., *Journal of Microelectromechanical Systems*, 12, 233–242, 2003.)

1. Detection, localization, and tracking of a transmitter of acoustic or electromagnetic energy, intruder, wild animal under observation, military troop movement, and submarine detection and tracking, and so on.

   Measurement of environmental parameters, such as temperature, humidity, rainfall, radioactivity, presence of dangerous gases, and so on, is readily carried out by a sensor network. Each measurement must be tagged with location information. An autonomous WSN can play the unique role of taking measurements in a hazardous environment, such as a reactor chamber, a furnace, a radioactive reactor, and so on.
2. Detection and localization information and the observed environmental parameters along with location information are required to be transmitted by the sensor nodes to a central information processor, which fuses the sensor information and transmits it to a more distant command center via the Internet. Since the available power at each sensor node is limited, we should choose the shortest multihop communication link connecting a node to the information receiver.

A WSN has many blind sensors randomly distributed around an object of interest (e.g., a forest fire). Location of all blind nodes has been previously estimated. There is one or more anchor nodes (previously used in the localization step) equipped with large battery power for transmission of data received from the neighboring blind nodes in a single hop or multiple hops. Anchors in turn communicate with other anchors and finally to the Internet via a gateway. The data is now made available to a distant computer for advanced processing. The overall structure of a typical WSN is shown in Figure 1.12. There are many possible distributions of nodes and anchors depending upon the specific application. Primary concerns in the design of a WSN are to reduce power consumption, reduce delay in transmission, increase reliability, and a few others as discussed in a recent review article [6]. Implementing energy efficient protocols may maximize network lifetime. Further, fault tolerance can be improved through a high level of redundancy by deploying more nodes than the minimum number required when all nodes are functioning properly. The transmission delay (latency) largely depends upon the network topology; it is low where nodes are connected to the anchor directly through a single hop, as in star topology; and it is high in large networks where two or more hops are required to reach the nearest anchor.

1.4 TYPES OF SENSORS

Sensors of common interest are those used for measurement of temperature, wind speed and direction, rain fall and soil humidity for agricultural applications, and water level in reservoirs or rivers for flood prediction. Another area of application is in forest fire prediction and control. Here, we basically need to measure the temperature, humidity, and

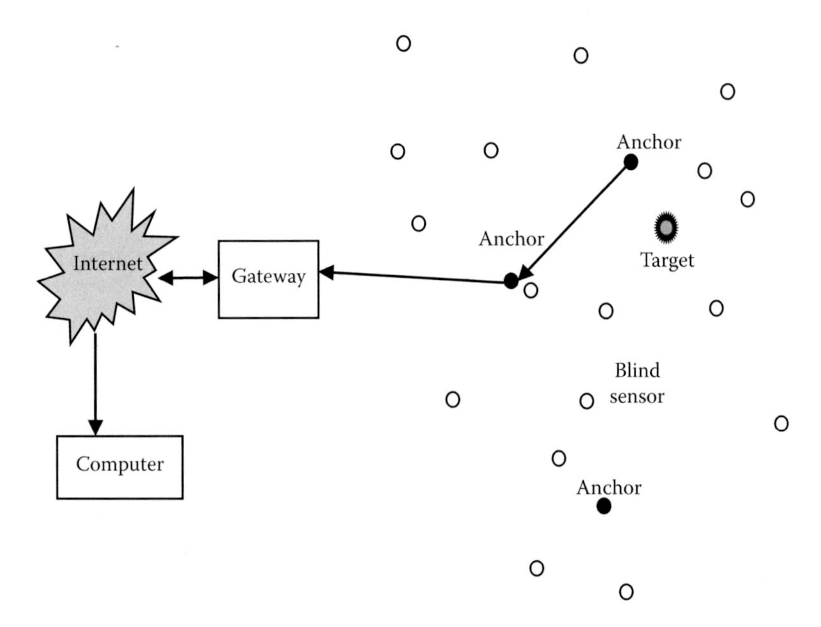

FIGURE 1.12 Wireless sensor network (WSN). Target may be any source of energy, such as a vehicle or an intruder (including a wild animal) or a forest fire.

wind speed and direction. Sensor arrays also find extensive application in security monitoring and movement of vehicles, including unmanned aerial vehicles (UAV) carrying material of strategic interest, underwater vessels, such as submarines, and unmanned underwater vehicles. Sensors for this type of work must be able to detect and measure sound and vibration both on ground and underwater. Finally, a radiation (infrared radiation) detector is used to detect warm bodies, such as humans and animals intruding into secure enclosures. Sensors differ from one another basically in the sensing element, that is, the mechanism to detect and measure a desired physical parameter; but other hardware, such as signal conditioners, A/D converters, data encoders, memory, transceivers, and so on, are common, except for their speed of operation. For example, a sensor for temperature monitoring is a low-speed device, as temperature variation is a slow phenomenon. It will consume less power compared with an acoustic sensor monitoring speech. All sensors are self-powered with a small button battery. The sensing element itself may not consume much power, but other electronic hardware accompanying the sensor consumes much of the power, particularly for transmission. Another common feature is that all electronic hardware is integrated into one or more chips, an application specific integrated circuit (ASIC), to minimize noise picked up by connectors and for compactness and robustness. Using modern micro-electromechanical system (MEMS) technology along with existing integrated circuit technology, it has become possible to build highly sensitive and miniature sensing devices.

An effective sensor array may have a few hundred sensors distributed over a wide geographic area. There are indeed tiny sensors, often nicknamed as dust particles, spread over a large area or volume where a target of interest is located. The task of the sensor array is to localize the target and to determine what should be transmitted. Sensors in the array may also have to localize themselves as they are often distributed randomly, as for example, when they are dropped from the air onto land or into ocean. Essential features of such sensors are

1. Light weight (<1 g) and small size (<5 mm^2).
2. Low power consumption (<1 μW).
3. Calibrated and stable output.
4. Inexpensive, since they are disposable.
5. Equipped to receive and transmit digital data.
6. Packaging should withstand environmental degradation, including mechanical impact and deep-water pressure.

A brief description of currently available sensors for different applications follows.

1.4.1 Thermal Sensor

Thermal sensors (e.g., thermocouple) utilize the Seebeck effect in which thermoelectric force is generated due to the temperature difference at the contact points between two different metals. The Seebeck coefficient mostly depends upon the composition of conductors. It may vary in the range of 100 μV/K to 1000 μV/K (K for Kelvin scale). A thermocouple is often used to measure high temperature, holding the cold junction at a known temperature. A thermopile is created by serially connecting thermocouples consisting of N + poly Si, P + poly Si, and Al (aluminum). A thermopile generates output voltages proportional to the local temperature differences. When connected serially, the voltage output between the first and last thermocouple is the sum of outputs of all thermocouples. Each thermocouple acts as a source of electromotive force (emf). By creating hot junctions on highly heat-resistant dielectric membranes, and cold junctions on highly heat-conductive silicon, it is possible to achieve a high-speed response and high-energy conversion efficiency.

Warm objects (e.g., human beings) radiate infrared radiation, which may also be used to measure the temperature of a warm object without contact. Omron has built a thermal sensor to measure infrared radiation and convert the radiation intensity into temperature. A silicon lens collects radiated heat in the far infrared region (wavelength of 4–14 μm) emitted from an object onto the thermopile sensor. The device is known as D6T, created entirely from Omron's proprietary MEMS, ASIC, and other application-specific parts to ensure high sensitivity and a high signal-to-noise ratio [19]. The range of detection is approximately 5–6 m. The temperature resolution as per the data sheet is 0.14°C. The device is, however, not yet ready for use outdoor and in hostile environments.

1.4.2 Microphone

The MEMS microphone is basically a variable capacitor, and its transduction principle is the change of coupled capacitance between a fixed plate (back plate) and a movable plate (membrane), caused by the incoming sound wave. An integrated circuit converts the change of the polarized MEMS capacitance into a digital (PDM modulated) or analog output according to the microphone type. The MEMS microphone is housed in a package with the sound inlet placed at the top or in the bottom part of the package; hence the top-port or bottom-port nomenclature of the package.

Features of the MEMS microphone (see Figure 1.13) made by ST Microelectronics [7] are summarized in Table 1.1. The sensitivity is the electrical signal at the microphone output for a given acoustic pressure at the input.

The sound pressure level (SPL), expressed in decibel, dBSPL = 20*Log(P/Po), where Po = 20 μPa is the threshold of hearing, and Pa stands for Pascal (a unit of pressure). 20*Log(1Pa/20 μPa) = 94 dBSPL. Thus, 1 Pa in dB is 94 dBSPL.

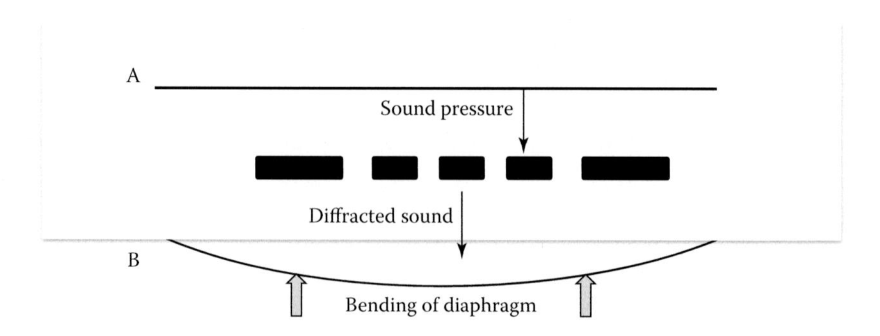

FIGURE 1.13 A schematic structure of MEMS microphone made by ST Microelectronics (USA). Incoming sound pressure buckles the diaphragm, thus changing the capacitance. The change in capacitance is a measure of sound pressure.

TABLE 1.1
Features of MEMS Microphones

Feature	Range
Sensitivity	−26 dBs (digital)
Directivity	Omnidirectional
SNR	61–63 dB (digital)
	63–68 dB (analog)
Maximum current	600–650 μA (digital)
Consumption	250 μA (analog)
Package dimensions	5×4×1–3×4×1 mm

Source: ST Microelectronics, Application Note AN4426, Tutorial for MEMS microphones, February 2017.

1.4.3 Hydrophones

MEMS technology and bionic principle are used to develop a low-frequency, high-sensitivity, three-dimensional omni-vector hydrophone that can obtain vector information of an underwater sound field by imitating the auditory principle of the lateral line of a fish [7]. The key sensing element is a four-beam microstructure and a rigid plastic cylinder (see Figure 1.14). The microstructure consists of four cantilever beams and a plastic cylinder mounted on the central block. On each beam, there are two piezoresistors whose resistance changes when deformed. The density of the plastic cylinder is close to that of the liquid in which it is immersed, and its radius (100 μm) is much smaller than the wavelength. Thus, the motion of the cylinder becomes very close to that of a fluid particle. As a sound wave propagates across

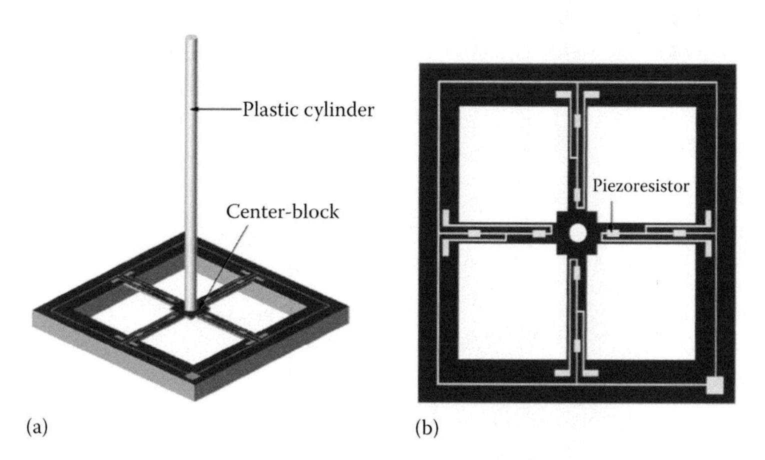

FIGURE 1.14 Microstructure (3D view) of a MEMS hydrophone consists of four beams attached to a central block, which also carries a rigid plastic rod. Two piezoresistors are mounted on each beam. The microstructure is immersed in castor oil (≈ 1 g/cm^3) enclosed in a polychloroprene rubber, which is transparent to low frequency sound. (From Xue, C., *Microelectronics Journal*, 38, 1021–1026, 2007.)

the microstructure sensor, the central block is subjected to a horizontal displacement and an angular rotation. This movement deforms the piezoresistors mounted on the beam. Change in the resistance of the piezoresistors is measured with help of a Wheatstone bridge. The rigid plastic cylinder mimics the role of stereocilia, and the piezoresistors mimic the role of hair cells in the lateral line of a fish.

The sensitivity of the device (Figure 1.15) as reported in [8] is –197 dB (Re 1v/μPa), and its resonance frequency in silicon oil is 350 Hz [9]. Experimental results show a flat frequency response, exhibit a sensitivity of –185 dB (x and y components) and –181 dB (z component) (kHz, 0dB = 1 V/μPa). At present, most two- and three-vector hydrophones are manufactured using the traditional combination approach of joining packaging of multiple accelerometers. This approach increases volume, power consumption and cost.

1.4.4 Accelerometer

One of the most common inertial sensors is the accelerometer, a sensor with a vast dynamic range. Accelerometers are available that can measure acceleration on one, two, or three orthogonal axes. They are typically used in one of three modes:

1. As an inertial sensor for measurement of velocity and position
2. As a sensor of inclination, tilt, or orientation in two or three dimensions, as referenced from the acceleration of gravity (1 g = 9.8 m/s^2)
3. As a vibration or impact (shock) sensor

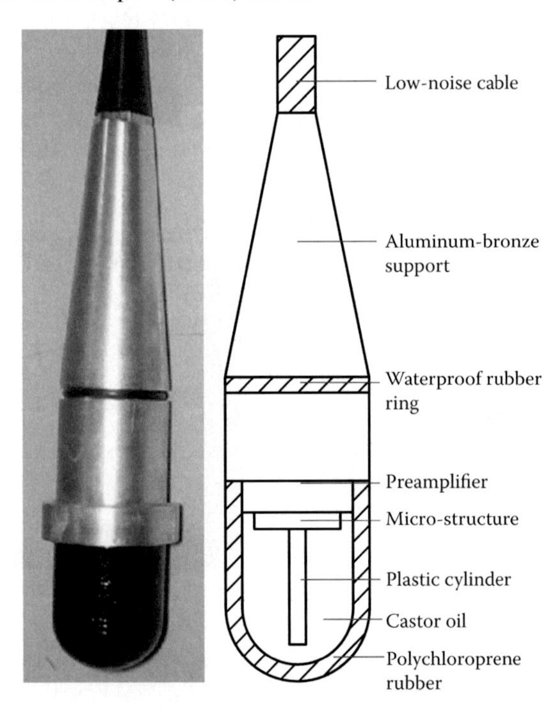

FIGURE 1.15 Photograph of packaged hydrophone. (From Xue, C., *Microelectronics Journal*, 38, 1021–1026, 2007.)

Most accelerometers are (MEMS). The basic principle of operation behind the MEMS accelerometer is the displacement of a small proof mass etched into the silicon surface of an integrated circuit (IC) and suspended by small beams. Consistent with Newton's second law of motion (**F** = **ma**), as acceleration is applied to the device, a force develops that displaces the mass. The support beams act as a spring, and the fluid (usually air) trapped inside the IC acts as a damper, resulting in a second order lumped physical system. The accelerometer has limited operational bandwidth and non-uniform frequency response [11].

There are several other principles upon which an analog accelerometer can be built. Two very common types utilize capacitive and piezoelectric effect to sense the displacement of the proof mass proportional to the applied acceleration (see Figure 1.16) The accelerometer that implements capacitive sensing outputs a voltage dependent on the distance between two planar surfaces. One or both "plates" are charged with an electrical current. Changing the gap between the plates changes the electrical capacity of the system, which can be measured as a voltage output. This method of capacitive sensing is known for its high accuracy and stability.

Capacitive accelerometers are also less sensitive to noise and variations of temperature, typically dissipate less power, and can have larger bandwidths due to internal feedback circuitry [11].

Analog devices pursued an integrated approach to MEMS where the sensor and signal conditioning electronics are on a single chip. The latest-generation accelerometer, ADXL363, is the result of more than a decade of experience in building

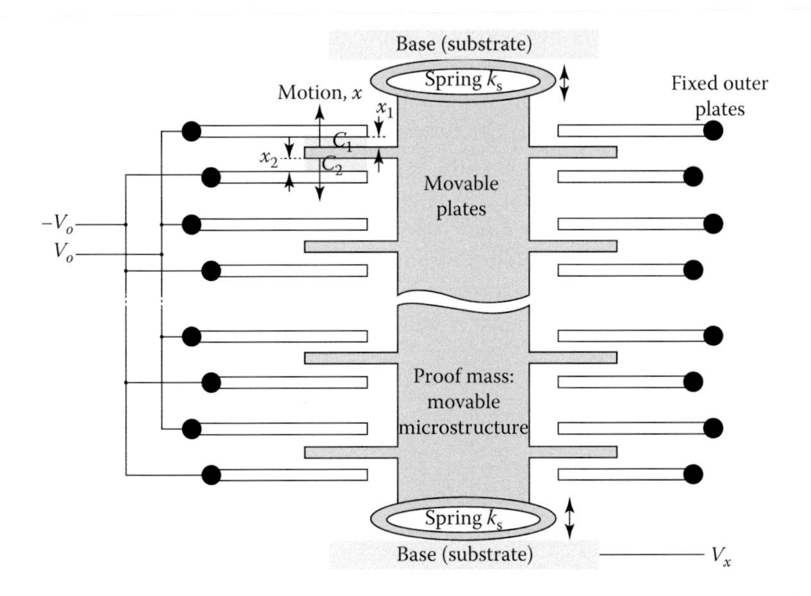

FIGURE 1.16 A simplified sketch of accelerometer structure. Proof mass is attached through springs at the substrata. When a force (acceleration) is applied to the proof mass, there is displacement proportional to the force. This displacement causes a change in the capacitance between movable and fixed plates. The change in capacitance is a measure of acceleration. (From Andrejasic M., MEMS Accelerometers, Seminar, Univ. of Ljubljana, 2008 [10].)

integrated MEMS accelerometers (Figure 1.17). The main features of the device are summarized as follows:

1. Accelerometer (three components) and temperature sensor
2. Ultralow power, 1.95 μA at 100 Hz output data rate
3. Twelve-bit resolution for all sensors (acceleration of 1 mg/LSB) and temperature scale factor of 0.065° C/LSB)
4. Adjustable threshold sleep/wake modes
5. Package measuring $3 \times 3.25 \times 1.06$ mm^3

By integrating the output of an accelerometer, we can get the velocity measurement just as in a coil-based seismometer or geophone. Coil-based geophones are a proven technology that has long provided the industry with rugged, cheap, and self-powered sensors.

However, with the requirement of more quantitative seismic, the need for lighter, broader-band and better-calibrated sensors is emerging. Land crews with large arrays find it difficult to set up and handle large collections of phones (e.g., 400,000+). In addition, with a renewed interest in multi-component recording, there is a need for new types of three-component receivers with a tight integration between field electronics and sensors. All these trends have led to the development of new digital sensors based on MEMS accelerometers. From the specification point of view, the essential benefit of MEMS accelerometers is a broadband linear amplitude and phase response that may extend from 0 (DC) to 800 Hz within ±1% in amplitude and ±20 μs in time (see Table 1.2). MEMS resonance frequency is far above the seismic band (<1 kHz). This makes it possible to record frequencies below 10 Hz without attenuation, including the direct current related to gravity acceleration. The gravity vector provides a useful reference for sensitivity calibration and tilt measurement [12].

Piezoelectric sensing of acceleration is natural, as acceleration is directly proportional to force. When certain types of crystal are compressed, charges of opposite polarity accumulate on opposite sides of the crystal. This is known as the piezoelectric effect. In a piezoelectric accelerometer, charge accumulates on the crystal and is translated and amplified into either an output current or voltage. But piezoelectric accelerometers only respond to AC phenomenon, such as vibration or shock. They have a wide dynamic range, but can be expensive

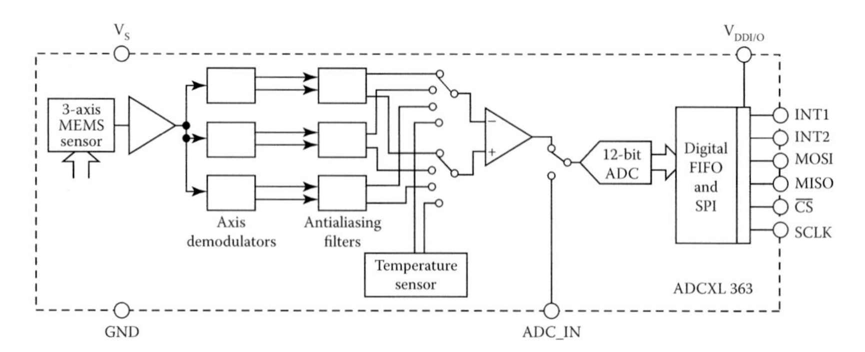

FIGURE 1.17 Functional block diagram of ADXL363 (ADXL363 Data sheet).

TABLE 1.2
Specifications of MEMS Accelerometer

Specification	Value
Size	Approx. 10 mm
Weight	About 1 g
Bandwidth	0 (DC)–800 Hz
Transmission	Fully digital

depending on their quality [13]. Piezo film based accelerometers are best as they are inexpensive, and respond to other stimuli, such as temperature, sound, and pressure.

1.4.5 Electromagnetic (EM) Sensors

Electric and/or magnetic fields are common attributes of targets with large internal currents. Examples of these include submarines, ships, trains, speeding cars, overhead HTDC cables [14], and so on. Distributed magnetic or electric sensors can be used to localize and measure the physical parameters of such targets. Here we emphasize basic principles used in the construction of sensors for measurement of electric and magnetic fields.

1.4.6 Electric Sensors

Force per unit charge exists in the presence of charges or charged bodies. An electric field may be static when charges do not move or move at a constant velocity. A magnetic field is produced when charges are moving in a conducting medium or in space. If currents vary in time, both an electric and a related magnetic field, called an electromagnetic field, is produced. Electric field sensors operate on the physical principles of an electric field and its effect, primarily its capacitance. Capacitance is the ratio between charge and the potential of a body (coulombs/volt $Q/V =$ Farad is unit of capacitance). Any two conducting bodies, regardless of their size and the distance between them, have a capacitance. The capacitance depends on various factors, for example, the distance between two charged plates (Figure 1.18), plates arranged in different configurations, the changing type and position of dielectric, the presence of any material in the neighborhood, and so on. The capacitance increases, indicating distance. Measuring the capacitance may sense all these parameters.

1.4.7 Magnetic Sensors

Sensors to detect/measure a magnetic field are largely based on the Hall effect or magneto-resistive effect. The Hall effect is present in all conductive materials, and it is particularly pronounced in semiconductors. Consider a wire carrying current I and a magnetic field B across the wire (Figure 1.19).

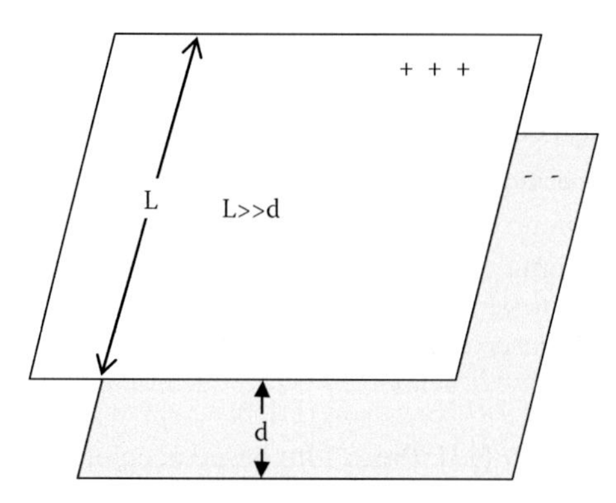

FIGURE 1.18 Parallel conducting plate capacitor. $C = \varepsilon_0 \varepsilon_r S/d$ where ε_0 is permittivity of vacuum, ε_r is relative permittivity (dielectric constant) of the intervening material, and S is the area of the plate ($S = L^2$). ε_r of some common materials are: nylon: 3.1, rubber: 3.0, glass: 6.0, distilled water: 81, and silicon: 11.

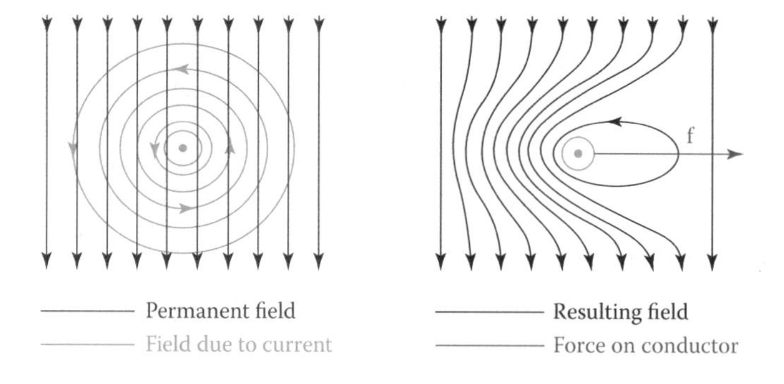

FIGURE 1.19 Mechanical force on current carrying conductor due to distortion of external magnetic line of force.

The magnetic field generated by the current (Ampere magnetic field) distorts lines of force due to the external magnetic field. As a result (Newton's third law), a force perpendicular to both current and magnetic field is established:

$$F = k_H I B_\perp$$

where $B_\perp$ is the component of a magnetic field perpendicular to the current vector and k_H is a coefficient dependent on the material (for copper, $k_H = -5.4 \times 10^{11}$).

1.4.8 Hall Effect

Consider a patch of conductor in place of a wire conductor. The Hall effect is based on the Lorentz force felt by charge carriers moving in a magnetic field. The schematic of the classical configuration is shown in Figure 1.20, where a thin patch of a conductor is placed in a magnetic field B. When current flows in the x direction, the

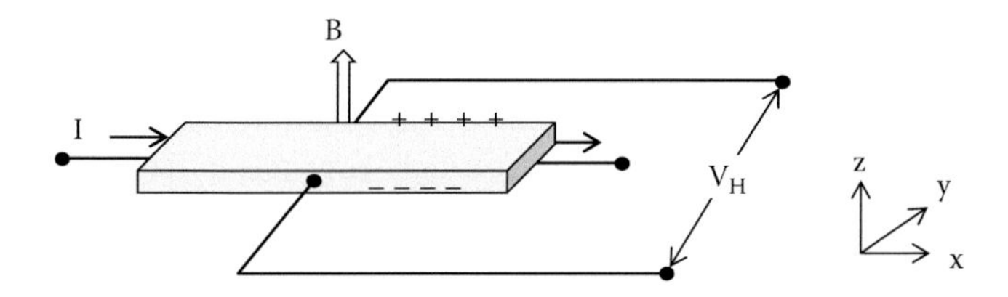

FIGURE 1.20 A thin slab of conductor is placed in a magnetic field. A force in the y-direction acts on electrons creating a charge. The Hall potential V_H is generated across the y-axis. The current is along the x-axis and the magnetic field along the z-axis.

Lorenz force acts in the y-direction, creating a charge distribution that counter balances the force. Therefore, a (Hall) voltage V_H builds, as shown in Figure 1.20 [15].

$$V_H = \frac{IB}{qnt}$$

where t is the thickness of the patch, n is the carrier charge density (charge/m^3), and q denotes the charge of the electron (measured in Coulombs). The Hall coefficient $k_H = 1/qnt$ is a material property.

Hall effect sensors are commonly fabricated by standard complementary metal-oxide semiconductor (CMOS) technology. In general, they can be used to measure a magnetic field in the range of 1 μT to 1 T (T: Tesla is Newton per ampere meter; 1 Tesla = 10,000 Gauss; as an example, the Earth's magnetic field is about half Gauss), and they have a die size less than one mm. Their robustness and simple fabrication process justify their use in hundreds of applications.

In addition to the Hall effect, the application of an external magnetic field causes an increase in resistance, often known as the magneto-resistive effect. The increase in resistance is due to a longer curved path followed by the electrons due to the Hall effect. The resistance of the device becomes a measure of the magnetic field. An approximate relation between increase in resistance ΔR and magnetic field B (component perpendicular to the strip) is given by

$$\frac{\Delta R}{R} = kB^2$$

where k is a constant depending on the material (e.g., Fe19Ni81 (Permalloy) k = 2.2).

1.4.9 Nanosensors

Finally, it is worth mentioning the upcoming nanosensors, whose sensing element is a carbon nanotube (CNT) or a graphene nanoribbon (GNR) working in the terahertz band (0.1–10.0 THz) (T for Tera 10^{12}). CNT is … Graphene is a one-atom-thick planar sheet of bonded carbon atoms in a honeycomb crystal lattice. CNT: A folded strip of graphene is GNR. It has been shown that for a maximum antenna size in the order of several hundred nanometers (the expected maximum size for a nanodevice), both a nanodipole and a nanopatch antenna can radiate electromagnetic waves in the terahertz band (0.1–10.0 THz) [16,17]. This is a very large bandwidth making it possible to transmit femtosecond-long (10^{-15} s) pulses. Having a huge bandwidth invites us to rethink not only the communication aspects of wireless networks, but also the networking issues. The potential impact of MEMS has prompted many efforts to commercialize a wide variety of novel MEMS products. In addition, MEMS are well suited for the needs of space exploration and thus will play an increasingly large role in future missions to the space station [18].

REFERENCES

1. C. U. Padmini and P. S. Naidu, Circular array and estimation of direction of arrival of a broadband source, *Signal Processing*, vol. 37, pp. 243–254, 1994.
2. P. S. Naidu, *Array Signal Processing*, 2nd edition, Boca Raton, FL: CRC Press, 2010.
3. L. M. P. Leao de Brito and L. M. R. Peralta, An analysis of localization problems and solutions in wireless sensor networks, *Polytechnical Studies Review*, vol. 8, no. 9, pp. 1–27, 2008.
4. B. A. Werneke and K. S. J. Pister, MEMS for distributed wireless sensor network. *Proceedings of 9th International Conf. on Electronics, Circuits and Systems*, vol. 1, Dubrovnik: IEEE, 291–294, 2002.
5. L. Zhou, J. M. Kahn, and K. S. J. Pister, Corner-cube retro reflectors based on structure-assisted assembly for free-space optical communication, *Journal of Microelectromechanical Systems*, vol. 12, pp. 233–242, 2003.
6. P. Rawat, K. D. Singh, H. Chaouchi, and J. M. Bonnin, Wireless sensor networks: Recent developments and potential synergies, *Journal of Supercomputing*, vol. 68, pp. 1–48, 2014.
7. ST Microelectronics, Application Note AN4426, Tutorial for MEMS microphones, February 2017.
8. C. Xue, S. Chen, W. Zhang, B. Zhang, G. Zhang, and H. Qiao, Design, fabrication and preliminary characterization of a novel MEMS bionic vector hydrophone, *Microelectronics Journal*, vol. 38, pp. 1021–1026, 2007.
9. G. Zhang, P. Zhao, and W. Zhang, Resonant frequency of silicon micro-structure of MEMS vector hydrophone fluid-structure interaction, *AIP Advances*, vol. 5, pp. 1–8, 2015.
10. M. Andrejasic, MEMS accelerometers, seminar, University of Ljubljana, 2008.
11. M. Elwenspoek and R. Wiegerink, *Mechanical Microsensors*, New York: Springer, 1993, pp. 132–145.
12. D. Mougenot, Digital accelerometer: How they have impacted the seismic industry in 10 years? 8th Biennial International Conference, Hyderabad, India, 2010.
13. J. Doscher, Accelerometer design and applications, company (analog devices) brochure, Norwood, MA, pp. 1–61, 2005.
14. Y. Cui, J. Lv, H. Yuan, L. Zhao, Y. Lui, and H. Yang, Development of a wireless sensor network for distributed measurement of total electric field under HVDC transmission lines, *International Journal of Distributed Sensor Networks*, vol. 2014, pp. 1–9, 2014.
15. M. T. Todaro, L. Sileo, and M. D. Vittorio, Magnetic field sensors based on micro electromechanical systems (MEMS) Technology, web publication. Available at www.Intechopen.com
16. I. F. Akyildiz and J. M. Jornet, Electromagnetic wireless nanosensor networks, *Nano Communication Networks*, vol. 12, pp. 3–19, 2010.
17. J. M. Jornet, I. F. Akyildiz, Graphene-based nano-antennas for electromagnetic nanocommunications in the terahertz band, Proceedings of the Fourth European Conference on Antennas and Propagation (EuCAP), 2010.
18. J. W. Judy, Microelectromechanical system (MEMS): Fabrication, design and applications, *Smart Materials and Structures*, vol. 10, pp. 1115–1134, 2001.
19. OMRON Corporation, High sensitivity enables detection of stationary human presence, data sheet of D6T.

2 Basic Inputs

The basic information required to locate a source, in two (plane) or three (volume) dimensions, is the measure of the distance from a set of reference points whose locations are known. There are two basic distance measurements: decline in strength of field (e.g., electric field) or power (e.g., acoustic power) and time of travel from transmitter to receiver. The inverse power law is the basis for the distance measure. The received signal strength (RSS) is widely used as the distance measure for localization with a distributed sensor array (DSA). RSS measurements are inexpensive and simple to implement in hardware. There is no need for time synchronization. Yet RSS measurements are notoriously unpredictable, largely because of multiple reflections and loss of power due to scattering. RSS measurements have been used for localization in indoor problems. Time of -arrival (ToA) and time difference of arrival (TDoA) are two other commonly used inputs for localization. If the source is mobile, frequency difference of arrival (FDoA) has been used in instantaneous localization. In addition to the previous inputs we describe four other methods, namely, the lighthouse effect, the direction-of-arrival (DOA) and the phase change in a multitone signal and instantaneous frequency.

2.1 RECEIVED SIGNAL STRENGTH

In distributed sensor arrays (DSAs), there are a few nodes known as anchor nodes where an omnidirectional sensor (receiver) measures signal strength. It is assumed that, in the near-field region, there is only one source (transmitter) active at a time. In a very simplified situation, with open space and no reflections, the signal strength at an anchor node due to a point source is inversely proportional to the square of the distance r (geometrical spreading),

$$P = \frac{s_0}{r^2}$$

where s_0 is the power radiated by the source. In a real channel, the RSS will decay much faster, often modeled as

$$P = \frac{s_0}{r^\alpha}$$

where α is a constant in the range of 2–4 [1]. Note that, in free space, $\alpha = 2$. Let P_0^{dB} be power in decibels measured at distance r_0. We can express the power measured at distance r in terms of P_0^{dB}, where P_0^{dB} is ($10\log(s_0)$) transmitter power in decibels.

$$P_r^{dB} = P_0^{dB} - 10\alpha \log \frac{r}{r_0}$$

We can also express r in terms of r_0

$$r = r_0 10^{(P_0^{dB} - P_r^{dB})/10\alpha} \tag{2.1}$$

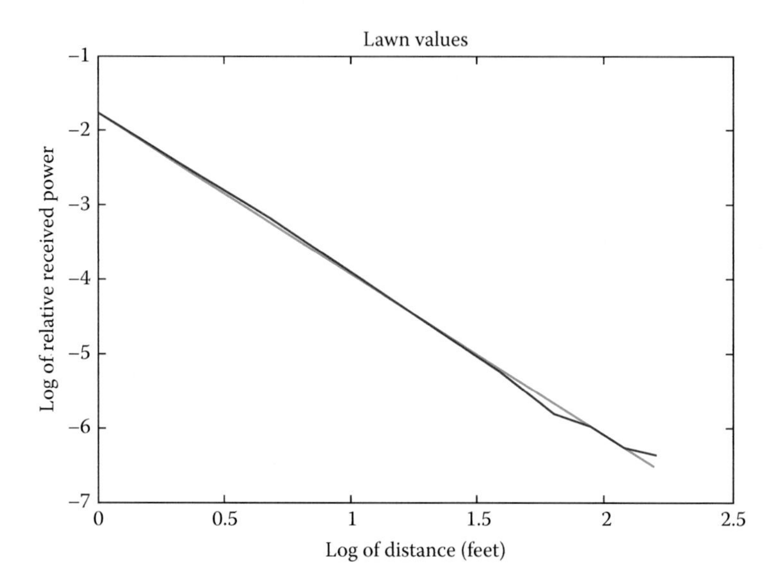

FIGURE 2.1 Received signal strength measurements in open space (lawn). We have plotted the logarithm of the received power (in arbitrary units) as a function of the logarithm of distance (in feet). Green curve: Least-squares linear fit to black curve. The slope of the line is $\alpha = 2.1656$.

Equation 2.1 enables us to measure the distance to all sensor nodes from the transmitter in units of distance to reference node r_0, which may be taken as 1 m. In this approach, there is no need to have control of the power emitted by the transmitter. But the effect of multipath propagation, where the line-of-sight (LOS) signal and the non-line-of-sight (NLOS) signals overlap, can severely limit the usefulness of the RSS method. NLOS and LOS signals are uncorrelated if the transmitted signal is made up of a pseudonoise (PN) sequence. Then, the received signal strength is just the sum of the powers of all multipath signals (no interference terms). The RSS method does not require any time synchronization. This is indeed a very useful property in a large distributed sensor network.

EXAMPLE 2.1

A simple experiment was carried out to measure parameter α, both in open space and in the laboratory corridor (the second floor under construction of the Electronic and Communication Engineering [ECE] Department). A pulsed sinusoidal waveform (100 Hz) was fed to a loudspeaker under computer control (laptop). A microphone, connected to another laptop, was used to receive the waveform and compute the received power. Both loudspeaker and microphone were aligned on the same axis and positioned 1 m above ground. The reflections from the ground were filtered out through time gating.

The received signal was sampled at a rate of 2000 samples/sec, giving 1000 signal samples (half a second in duration). The average power over the signal duration was computed and plotted as a function of a logarithm of distance. The value of α is determined by the slope of the line that provides the best straight-line hit to the plot (see Figure 2.1). A group of final-year engineering students led by Ms. Vysalini carried out the previously mentioned experiment.

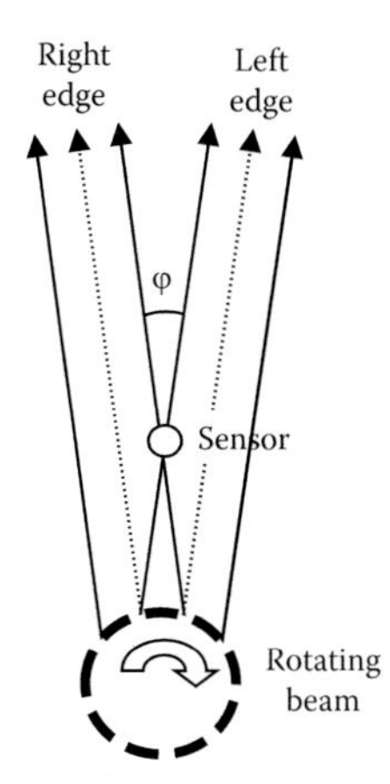

FIGURE 2.2 A sensor is illuminated with a parallel beam of width b. The actual duration of time the sensor is illuminated depends upon the radial distance from the axis of rotation.

2.2 LIGHTHOUSE EFFECT

Here is a simple method of ranging [2]. A rotating parallel beam of light illuminates stationary sensors, each of which is equipped with a light-detecting sensor. We measure the duration of the time that a sensor remains under illumination. The time duration is inversely proportional to the d range. Consider a parallel beam of light of width b ($\ll d$) produced by an optical instrument, which rotates in the plane of sensors at a constant angular speed (see Figure 2.2). A sensor enters the beam of light on the right edge and exits on the left edge. The sensor will traverse an arc of length approximately equal to b (beam width). This arc will subtend an angle φ (radians) at the center of rotation. This may be approximately given by

$$\varphi \approx \frac{b}{d}$$

where d is distance to the center of rotation. Let ω be the angular speed of rotation (radians/sec). The duration of time during which the sensor is under illumination is

$$\tau = \frac{\varphi}{\omega} = \frac{b}{d\omega} \tag{2.2a}$$

From Equation 2.2a, we can estimate the distance, given the duration of τ, as

$$d = \frac{b}{\tau\omega} \tag{2.2b}$$

Thus, the duration of illumination is inversely proportional to the distance. As in the RSS method, each sensor can autonomously determine its range from the transmitter. Here, too, there is no need for time synchronization.

The most critical requirement in this approach is the formation of exact parallel beams with the help of a precision optical system and laser light source. The use of other types of radiation, such as ultrasonic [3] or ultra-high frequency (UHF) electromagnetic radiations [4], has been explored. The methods seem to be well suited for a small distributed sensor network without a requirement of clock synchronization, as in other methods based on measurement of time. We describe here the actual implementation of the algorithm.

2.2.1 Beam Generation

Roemer [1] suggested a simple method to generate a parallel beam. He used two solid-state lasers mounted on a slowly rotating platform. A laser beam is incident on a tiny mirror held at an angle of 45° so that the reflected beam is at ⊥ to the incident beam. The setup is shown in Figure 2.3. Two lasers are used to produce the outline of a laser beam (Figure 2.3a). The beam will be perfectly parallel when the mirrors are positioned exactly at 45°, which is not an easy task! The mirrors are rotated about the x-axis at high speed in comparison to the rotational speed of the platform on which the entire optical system is mounted. The platform is in the x–y plane and turns around the z-axis (see Figure 2.3a).

The optical beam lies perpendicular to the plane of paper, that is, in the y–z plane. A target on the x-axis is intercepted by beam outlines at a time interval depending on the angle that the arch subtends at the center of rotation. Assume the platform is turning clockwise; the lower beam edge will intercept the sensor first, followed by the upper edge, the time interval between the interceptions being proportional to the length of the arc through the sensor, as shown in Figure 2.4. This information is converted into the radial distance to the sensor. The distance to a sensor is inversely proportional to the angle subtended by an arc through points where the beam meets the sensor (Figure 2.4). The sensor is assumed in the plane of platform rotation (that is, $z=0$ plane). The elapsed time between two encounters with the sensor, divided by the actual time of full rotation, gives the angle φ. Equation 2.2b gives the distance from the axis of rotation in terms of elapsed time, τ, which is measured by a sensor.

Consider a situation where the sensor is above (or below) the plane of platform rotation. Then, the scanning beam would miss the sensor unless it is expanded in the z-direction as shown in the side view (Figure 2.3b). Ideally, in place of a sharp edge, we need an edge expanded in the y-direction. Then the beam will definitely encounter the sensor even though it is displaced along the z-axis. The distance is measured exactly as before. Since the beam has cylindrical l about the axis of platform rotation, all measured distance is always from the axis of rotation and is equal to the x-coordinate of the sensor. To obtain the y-coordinate, an identical system, except with the platform turning around the x-axis, is required. It is possible to measure all three coordinates with three independent platforms. This contrasts an arrangement where a parallel beam is produced with the help of a rotating mirror from a point source of light the distance measured from the center of rotations, which is simply a radial distance to the sensor. We will require at least three independent measurements to localize a sensor (known as triangular lateration). The advantage of cylindrical lateration is that we need just two orthogonal rotating platforms placed at the base station. The x-, y-coordinates of all sensors (lying in the x–y plane) can be measured directly without resorting to any complex mathematical analysis. Evidently this is of great importance when we have a large number of sensors with limited access to power.

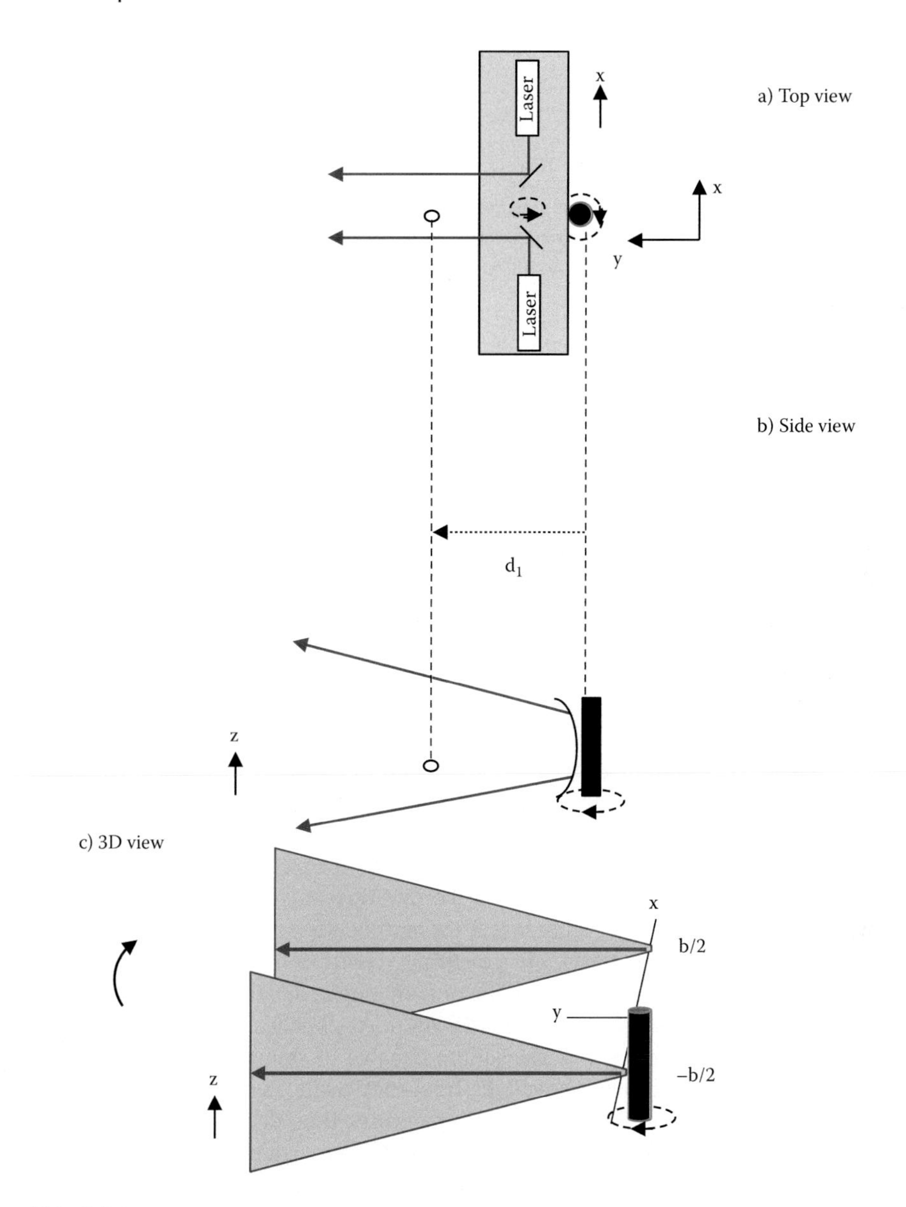

FIGURE 2.3 Principle of cylindrical lateration: (a) Two laser beams are incident on two tiny mirrors placed at 45°. The reflected laser beams produce an outline of a wide parallel beam in (x, y) plane (z = 0). (b) The beam is expanded in the z direction. (c) 3D view of the scanning beam. As the platform turns around the z-axis, much of the upper part of the x–y plane is scanned.

The computational complexity is extremely low for the detection of two light pulses and for measuring the time interval between them. The interval time is then relayed to the base station directly. Furthermore, as mentioned earlier, there is no need for any time synchronization. Thus, each sensor is truly autonomous. Since on-board electronics has been reduced to the minimum, the power consumption is also low.

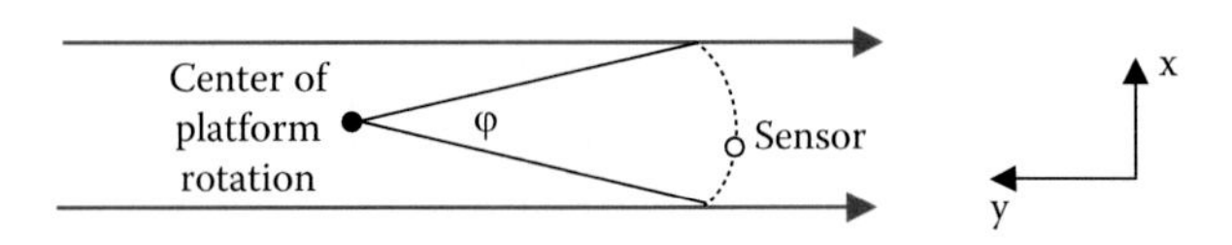

FIGURE 2.4 Assume the beam is turning clockwise; then the lower edge encounters the sensor first, followed by the upper edge. Angle φ is the measure of distance to the sensor. When the beam width is small compared with the range, we can use Equation 2.2b to estimate the range.

Power is required only to communicate on demand with the base station. The main limitation is the need for LOS connectivity between the base station and the sensor.

2.2.2 Prototype

A laboratory model to demonstrate feasibility has been built by Roemer [2]. A picture of the working model of the base station and sensor is shown [2]. An accuracy of 2% is claimed in a room-scale experiment.

2.2.3 Events with Large Propagation Speed Difference

When a source is simultaneously radiating an event with a very large speed difference, the difference in ToA can be used to estimate the distance of the source from the receiver. We assume LOS propagation. A familiar example is lightning and thunder. A light pulse from lightning travels about 10,000 times faster than the sound from thunder. The light pulse reaches all sensors instantly, but the sound pulse takes much longer depending on the relative distance to the lightning source. This fact has indeed been used to localize lightning [5]. This also forms the basis for earthquake range estimation. The time difference between the first arrival (p-wave) and later arrival (s-wave) is used for the range estimation of the earthquake epicenter [6]. This idea has also been exploited for localization in a distributed (wireless) sensor array [7].

A large number of tiny sensors equipped with a light signal sensor (i.e., photodiode) and an acoustic signal sensor (i.e., MEMS microphone) are distributed over an area of tens of square meters. A transmitter capable of simultaneously radiating a light pulse and coded acoustic signal (i.e., frequency modulated signal) is positioned at an unknown location. Each tiny sensor measures time delay and computes the range (time difference divided by speed of sound). Here again, there is no need for time synchronization. The range estimate is communicated back to the transmitter using a special optical device called a retroreflector. The previously mentioned experiment is repeated at many locations. All range information is pooled together and an algorithm similar to multidimensional scaling (MDS), described in detail in Chapter 6 is used to estimate the sensors as well as transmitters.

2.3 TIME OF ARRIVAL

ToA is a popular measure for distance estimationa. Given that propagation speed c is constant, the ToA is simply converted into distance as $c\tau$. The assumption of constant speed, however, may not be valid in some situations, such as electromagnetic

(EM) waves in a global positioning system (GPS) traveling though the ionosphere or seismic waves traveling through the Earth's crust. Supposing there are N sensors, there will be N ToA measurements, which we represent in a vector form

$$\boldsymbol{\tau} = \{\tau_1, \tau_2, \cdots \tau_N\}^T$$

ToA measurements are meaningful only when the transmitter broadcasts the exact time of transmission and the clocks at both the transmitter and receiver are synchronized. Alternatively, another synchronized sensor is placed close to the transmitter for timing information. This requires that the transmitter be a friendly device, that is, not controlled by an enemy agent. The clock function at each sensor is synchronized with a master clock installed at the central processing center or at the anchor node with the GPS receiver. A clock built in a sensor is necessarily an inexpensive device; therefore, it tends to drift with time. Attempts have been made to model the drift with an intention of possible correction. In general, clock function of the ith node is modeled as

$$C_i(t) = f_i t + \theta_i \tag{2.3}$$

where f_i is the clock skew frequency (which is the ratio of the actual frequency and the nominal frequency) and θ_i is the clock offset of ith node. Both parameters are constants over a short time interval. For an ideal clock, $f_i = 1$ and $\theta_i = 0$. Because of imperfections in the crystal oscillator commonly used in distributed arrays, the frequency of a clock varies up to 40 ppm, which means clocks of different sensors can drift as much as 40 μs in one second. In terms of error in distance estimation, it is 40 mm when the sensors are 1 km apart. A graphical illustration of the clock function of an inexpensive clock in a node and the clock function of an ideal clock, perhaps present in an anchor node with a GPS receiver, is shown in Figure 2.5. Clock synchronization is a procedure for providing a common time across a DSA (see Chapter 6 for the estimation of skew and offset). Synchronization is essential in a number of applications; in particular, it is of interest to localization. ToA and TDoA, essential pieces of information for distance measurement, involve time or relative time measurements. To map a sensor clock output into time reading of the reference sensor, we can employ Equation 2.3 provided we have an estimate of the parameters f (skew frequency) and θ (clock offset). Considerable effort has gone into the estimation of these two parameters.

Synchronization happens to be an important task in many applications of sensor array [8]. Here, we outline a method potentially useful in DSAs.

In one-way message dissemination, a reference node (i.e., anchor node) broadcasts its timing information to all nodes within communication range. Each node records the arrival time. The reference node broadcasts N messages at different time instants, $t_{ref,1}, t_{ref,2}, \cdots, t_{ref,N}$. The timing messages reach a node at local time instants, $t_{rec,1}, t_{rec,2}, \cdots, t_{rec,N}$. This message dissemination is illustrated in Figure 2.6. The time of arrival of a timing message at a selected node, in terms of the time of broadcast, is modeled as

$$t_{node,n} = f(t_{ref,n} + \tau + X_n) + \theta \tag{2.4}$$

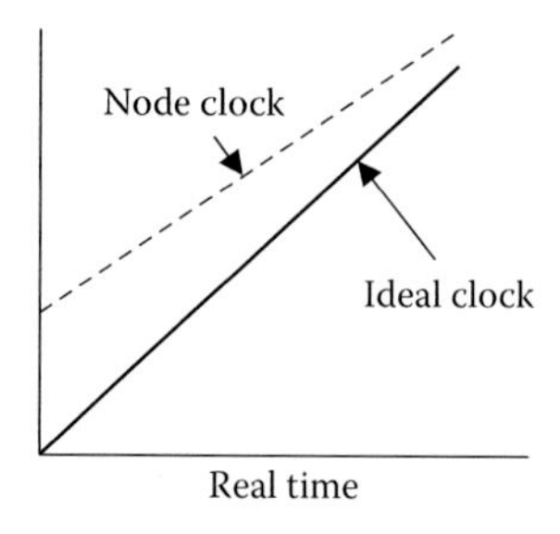

FIGURE 2.5 An illustration of clock function. Thick line stands for an ideal clock ($f = 1$ and $\theta = 0$) and the dashed line for an ordinary node clock ($f < 1$ and $\theta > 0$).

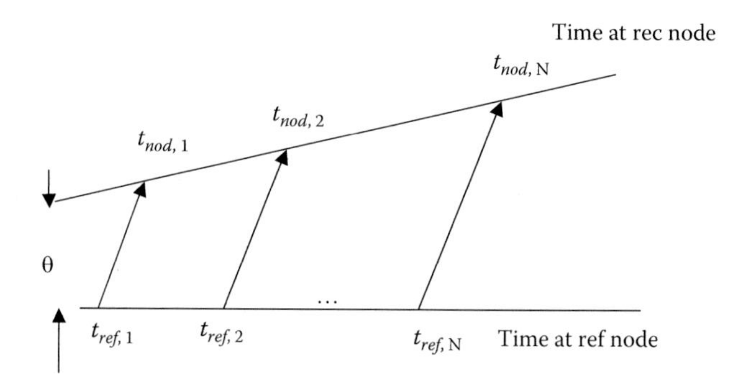

FIGURE 2.6 Illustration of one-way message dissemination.

where τ is constant delay (i.e., travel time from the reference node to the receiving node) and X_n is the variable delay in transmission, presumed to be random and uncorrelated over n. Since $f\tau$ is a constant, it may be combined with the clock offset. A modified clock offset is defined as $\theta' = f\tau + \theta$. Collecting arrival times for N rounds of transmission and expressing them in a matrix form, we obtain the following matrix equation (Figure 2.6):

$$\begin{bmatrix} t_{ref,1} \\ t_{ref,2} \\ \vdots \\ t_{ref,N} \end{bmatrix} = \begin{bmatrix} t_{node,1} & -1 \\ t_{node,2} & -1 \\ \vdots & \\ t_{node,N} & -1 \end{bmatrix} \begin{bmatrix} 1/f \\ \theta'/f \end{bmatrix} - \begin{bmatrix} X_i \\ X_2 \\ \vdots \\ X_N \end{bmatrix} \tag{2.5}$$

$$\mathbf{T}_{ref} \quad = \quad \mathbf{T}_{node} \quad \mathbf{q} \quad - \mathbf{X}$$

The least-squares solution of Equation 2.5 is given by

$$\hat{\mathbf{q}} = (\mathbf{T}_{node}{}^T \mathbf{T}_{node})^{-1} \mathbf{T}_{node}{}^T \mathbf{T}_{ref}$$

Having obtained the model parameters (f,θ'), we can compute the time at the reference node (or anchor node) within an error set by the variable delay component.

From Equation 2.4, we have the following formula to compute the time at the reference node:

$$\hat{t}_{ref,n} = \frac{t_{node,n} - \hat{\theta}'}{\hat{f}} - X_n$$
$$= [t_{node,n} - 1]\hat{\mathbf{q}} - X_n \tag{2.6}$$

where ^ refers to estimated quantity.

It is required that the path followed by the signal is the direct straight-line path from source to receiver. However, the received signal may also have traveled via one reflector or more reflectors or scattered as in a multipath environment. The direct path signal generally arrives first and it is the strongest compared with all other multipath signals. We assume that the previously mentioned requirements are satisfied.

2.4 TIME DIFFERENCE OF ARRIVAL

When a transmitter is beyond our control or lacks a synchronized clock for timing information, we consider relative ToA with respect to a reference sensor or between all possible pairs of the sensors. It is then necessary that all pairs of sensors be mutually synchronized. Suppose there are N sensors and the first sensor is the reference sensor. We will have N–1, TDoA measurements, $\tau_{2,1}, \tau_{3,1}, \cdots \tau_{N-1,1}$, where

$$\tau_{i,1} = \tau_i - \tau_1, \quad i = 2, 3, \cdots N$$

By choosing another reference, say, the fifth sensor, we get another set of N–1 measurements,

$$\tau_{i,5} = \tau_i - \tau_5, \quad i = 1, 2, 3, 4, 6, \cdots N$$

But it is easy to show that these measurements can be expressed in terms of those obtained with reference to the first sensor as shown here

$$\tau_{i,5} = \tau_i - \tau_5 = \tau_i - \tau_1 - (\tau_5 - \tau_1)$$
$$= \tau_{i,1} - \tau_{5,1} \tag{2.7}$$

This equivalence will fail in the event of measurement errors. Let $\eta_i, i = 1, 2, ...N$ be the measurement errors at N sensors. TDoA between ith and fifth in the presence of errors in ToA measurement errors is given by

$$\tau_{i,5} = \tau_i + \eta_i - \tau_5 - \eta_5$$
$$= \tau_i - \tau_1 - (\tau_5 - \tau_1) + \eta_i - \eta_5 \tag{2.8}$$
$$= \tau_{i,1} - \tau_{5,1} + \eta_i - \eta_5$$

Clearly, the equivalence relation given in Equation 2.7 no longer holds. There is an unknown additional noise term. Hence, we will stand to gain if we consider all possible pairs of TDoA measurements.

2.4.1 TDoA with a Reference Transmitter

In the previous description, we found the difference of arrival time at two synchronized nodes from a single common source. We can consider a situation where we have two sources, for example, a single moving source as in [9] or two sources of which one is at a known location. Observe times of arrival at an unsynchronized node. Let $t_{n,\,known}$ be the time of arrival from the known transmitter at the *n*th node as shown by its own clock, and let $t_{n,\,unknown}$ be the time of arrival from the unknown transmitter at the same node.

$$t_{n,known} = \tau_{n,known} + x_n$$

$$t_{n,unknown} = \tau_{n,unknown} + x_n$$

where x_n is the synchronization error at the *n*th node. The TDoA at the *n*th node now becomes independent of the synchronization error

$$\Delta t_n = t_{n,unknown} - t_{n,known} = \tau_{n,unknown} - \tau_{n,known}$$

Cancellation of the synchronization error is illustrated in Figure 2.7.

With a little modification, we can use the algorithm (see Chapter 3) meant for synchronized nodes to estimate the location of the unknown source. There is *no need* to synchronize all nodes. This is indeed a great relief in the implementation of DSAs.

A large reflector in place of a second transmitter will serve the same purpose, provided the reflector produces a clear identifiable signal.

2.5 FREQUENCY DIFFERENCE OF ARRIVAL

When a transmitter or sensor or both are in motion, there is a Doppler shift in the received signal. It is assumed that the transmitter radiates a narrowband signal with a center frequency f_c. The Doppler shift depends upon the relative velocity of the transmitter in direction of signal arrival. It also depends upon the speed of wave propagation, c. Assume that the sensor is stationary (a case of interest to us here). Let the transmitter velocity be a vector $\mathbf{u}_s = \{u_x, u_y, u_z\}$ and let the direction of signal arrival at *m*th sensor be given by its direction cosines:

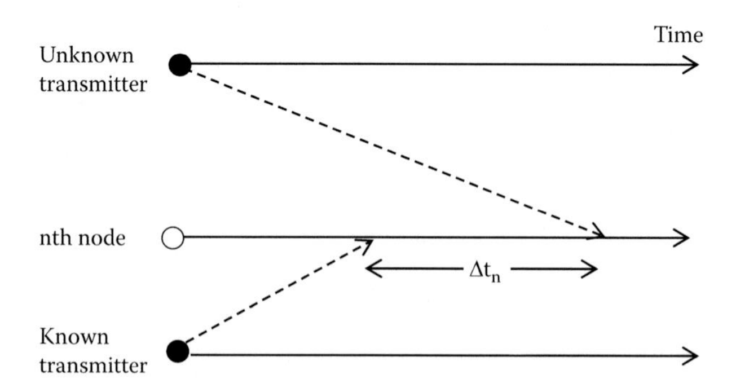

FIGURE 2.7 TDoA at a single node from two transmitters. The location of one transmitter is known. Synchronization errors get cancelled. There is no need for synchronization at all nodes in the array.

$$\{\sin\theta_m \sin\phi_m, \quad \sin\theta_m \cos\phi_m, \quad \cos\theta_m\}$$

where θ_m is the declination measured with respect to the z-axis and ϕ_m is the azimuth measured with respect to the y-axis in the x–y plane (see Figure 2.8). In terms of position vector $\mathbf{p}_s$, the velocity vector of the transmitter is

$$\mathbf{u}_s = \frac{d\mathbf{p}_s}{dt} = \begin{Bmatrix} \frac{dx_s}{dt} \\ \frac{dy_s}{dt} \\ \frac{dz_s}{dt} \end{Bmatrix}$$

and the Doppler shift is given by

$$\begin{aligned} c\frac{\delta f_m}{f_c} &= \frac{\mathbf{r}_{sm}^T}{|\mathbf{r}_{sm}|}\mathbf{u}_s \\ &= \{\sin\theta_m \sin\phi_m, \quad \sin\theta_m \cos\phi_m, \quad \cos\theta_m\}\mathbf{u}_s \end{aligned} \tag{2.9a}$$

where δf_m is the Doppler shift observed at the mth sensor. We can also relate the Doppler shift to the time derivative of range (range rate)

$$\begin{aligned} \frac{dr_{sm}}{dt} &= \frac{\partial r_{sm}}{\partial x_s}\frac{dx_s}{dt} + \frac{\partial r_{sm}}{\partial y_s}\frac{dy_s}{dt} + \frac{\partial r_{sm}}{\partial z_s}\frac{dz_s}{dt} \\ &= [\frac{\partial r_{sm}}{\partial x_s}, \quad \frac{\partial r_{sm}}{\partial y_s}, \quad \frac{\partial r_{sm}}{\partial z_s}]\mathbf{u}_s \\ &= \{\sin\theta_m \sin\phi_m, \quad \sin\theta_m \cos\phi_m, \quad \cos\theta_m\}\mathbf{u}_s \\ &= c\frac{\delta f_m}{f_c} \end{aligned} \tag{2.9b}$$

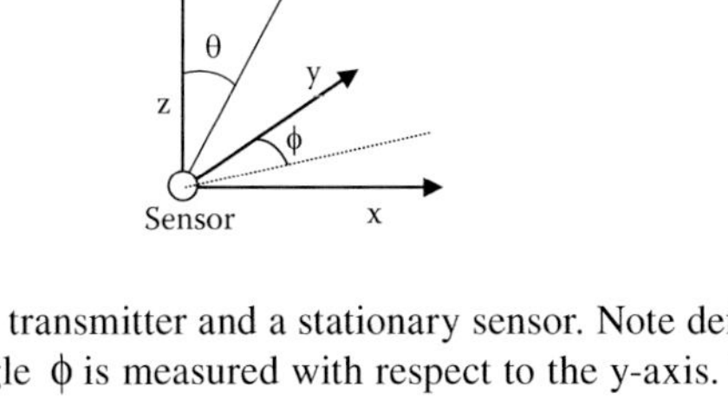

FIGURE 2.8 Moving transmitter and a stationary sensor. Note definition of angle of declination θ. Azimuth angle ϕ is measured with respect to the y-axis. We use this convention throughout this book.

Let us call $\delta f_m / f_c$ as the relative Doppler shift, which may be expressed as

$$\frac{\delta f_m}{f_c} = \begin{Bmatrix} \sin\theta_m \sin\phi_m \sin\theta_s \sin\phi_s + \\ \sin\theta_m \cos\phi_m \sin\theta_s \cos\phi_s + \cos\theta_m \cos\theta_s \end{Bmatrix} \frac{|\mathbf{u}_s|}{c} \tag{2.9c}$$

where θ_s is the declination (measured with respect to the z-axis) and ϕ_s is the azimuth (measured with respect to the y-axis) of the moving transmitter (Figure 2.8). Both θ_s and ϕ_s may be functions of time. We can write θ_m and ϕ_m in terms of the sensor and transmitter locations. Let the sensor (fixed) be at a point $(x, y, 0)$ and the transmitter (moving) be at $(x_s(t), y_s(t), z_s(t))$ at time instant t.

$$\sin\theta = \sqrt{\frac{(x_s(t)-x)^2 + (y_s(t)-y)^2}{(x_s(t)-x)^2 + (y_s(t)-y)^2 + z_s^2(t)}}$$

$$\sin\phi = \frac{(x_s(t)-x)}{\sqrt{(x_s(t)-x)^2 + (y_s(t)-y)^2}}$$

Using previous equations in Equation 2.9a, we obtain a simpler expression for the relative Doppler shift $\delta f_m / f_c$ at sensor m,

$$\frac{\delta f_m}{f_c} = \left(\frac{x_s(t) - x_m}{r_{sm}} u_s + \frac{y_s(t) - y_m}{r_{sm}} u_y + \frac{z_s(t)}{r_{sm}} u_z \right) / c \tag{2.9d}$$

We have the computed relative Doppler shift over an array of sensors located along the x-axis for a transmitter at a height of 100 m above the array and moving parallel to the x-axis with a speed of 10 m/sec. The results are shown in Figure 2.9a. Consider another possibility where the transmitter is moving at a constant speed parallel to the z-axis. The Doppler shift turns out to be a constant as seen in Figure 2.9b for motion along the z-axis.

The Doppler frequency difference between mth sensor and the reference sensor (say, sensor #1) is

$$\begin{aligned} \frac{c}{f_c}(\delta f_m - \delta f_1) &= \frac{c}{f_c} \Delta f_{m1} \\ &= \left(\frac{\mathbf{r}_{sm}}{|\mathbf{r}_{sm}|} - \frac{\mathbf{r}_{s1}}{|\mathbf{r}_{s1}|}\right)^T \mathbf{u}_s \end{aligned} \tag{2.10}$$

The FDoA is observable at each sensor with respect to a reference sensor, thus we have M-1 observations. But there are six unknowns (three spatial coordinates and three velocity components). Thus, we need at least seven or more sensors to estimate all six unknowns from the FDoA measurements. As in the case of RSS and lighthouse methods, it is not necessary for clock synchronization. Each sensor can act autonomously without having to wait for the time synchronizing signal. If the

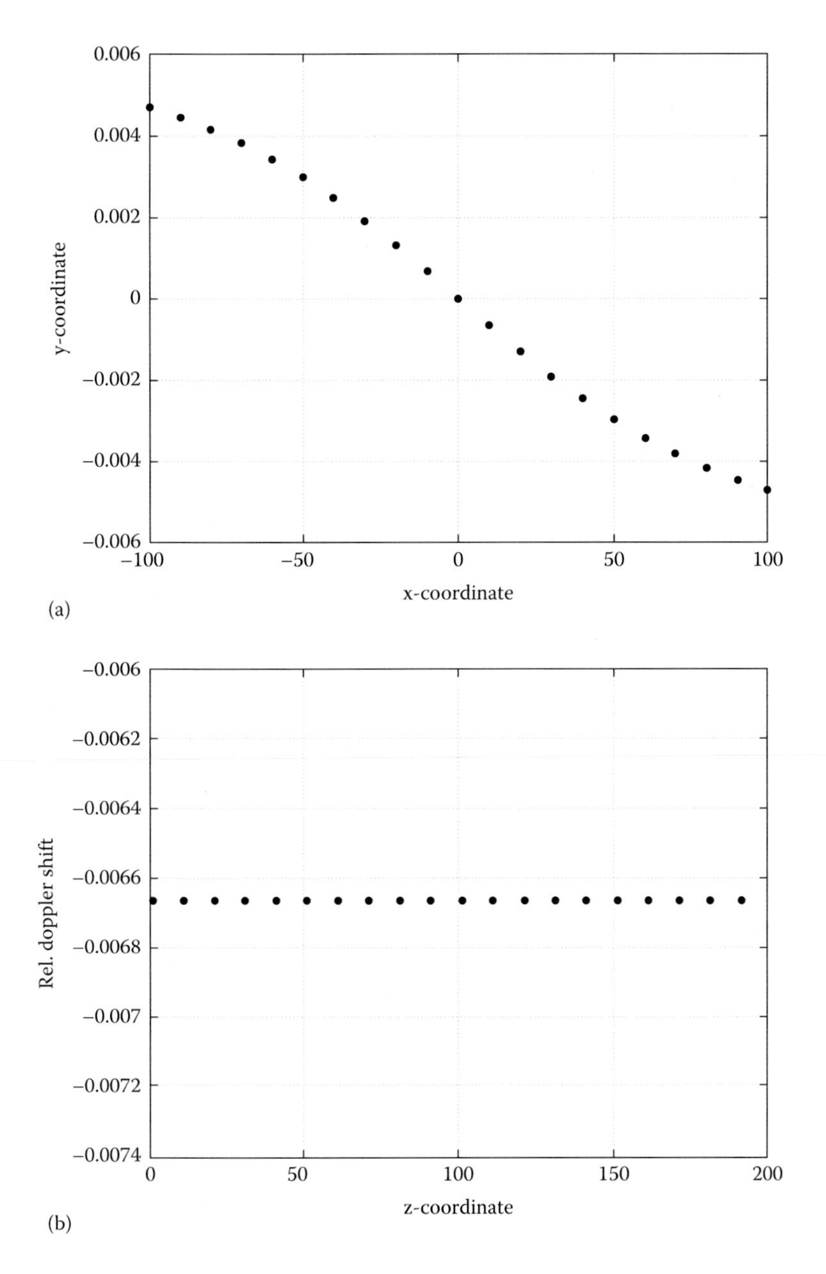

FIGURE 2.9 Relative Doppler shift (y-axis) as a function of transmitter position (in meters). (a) A transmitter is assumed to be moving along the x-axis at a height of 100 m with a speed of 10 m/sec in sea water (c = 1500 m/sec) and (b) The transmitter motion is along the z-axis. The relative Doppler shift remains constant.

transmitter signal is a pure sinusoid of known frequency, it is possible to estimate the frequency of the received signal at every sensor of the distributed array. This information is then communicated to the central processor (e.g., anchor node), which has a greater computing power for tracking a moving transmitter. One possible drawback of this method would be the need for a large observation interval to measure small Doppler shifts. We examine some of these issues in Chapter 3.

2.6 DIRECTION OF ARRIVAL

Estimation of the DoA of a signal may be obtained with the help of an array of sensors located at each anchor node. This may not be possible when the sensors are arbitrarily deployed, as in an autonomous sensor array. However, a cluster of sensors around an anchor may be used to form a beam and thus obtain DoA information. The subject of DoA estimation is beyond the scope of this work. It is well covered in several publications, for example, in [10,11]. Here, we stress that randomly distributed sensors around an anchor node and within communication range may be used for DoA estimation. As an example, we consider a cluster of 16 sensors randomly distributed within a radius of 10 m (communication range is assumed at 10 m), as shown in Figure 2.10a. The incident wave is a plane wave of wavelength 0.1 m. The array response (magnitude square) shown in Figure 2.10b, exhibits a significant variability away from the position of peak. By a simple process of averaging over the adjacent frequency bins, we can smooth out much of this scatter. The averaged response is shown in Figure 2.10c.

The DoA information may be combined with ToA information for transmitter localization. DoA provides an additional constraint on source location through the following relation:

$$\tan\theta_{DoA} = \frac{x_a - x_s}{y_a - y_s} \tag{2.11}$$

where θ_{DoA} is the DoA, (x_a, y_a) and (x_s, y_s) are coordinates of anchor node and transmitter location, respectively. The anchor locations (x_a, y_a) and Equation 2.11 act as linear constraints on coordinates of the unknown transmitter, x_s and y_s.

$$y_s = \frac{x_s}{\tan\theta_{DoA}} - \frac{x_a - \tan\theta_{DoA} y_a}{\tan\theta_{DoA}} \tag{2.12}$$

Equation 2.12 represents a linear relation between x_s and y_s coordinates of the unknown transmitter. If we have one more anchor, it will be possible to get one more similar constraint.

The intersection of these two linear lines will yield the unknown transmitter position.

Let us plot two linear constrains in the (x_s, y_s) plane. The two lines will intersect at a point where the transmitter is actually located. Thus, it is possible to localize a source just with two anchor nodes, along with a set of randomly distributed sensors, all of which lie within the communication distance to either anchor.

As an example, we demonstrate in Figure 2.11, given two correct estimations of DoAs, linear constraints defined by Equation 2.11 the intersect at a point where the source is actually located.

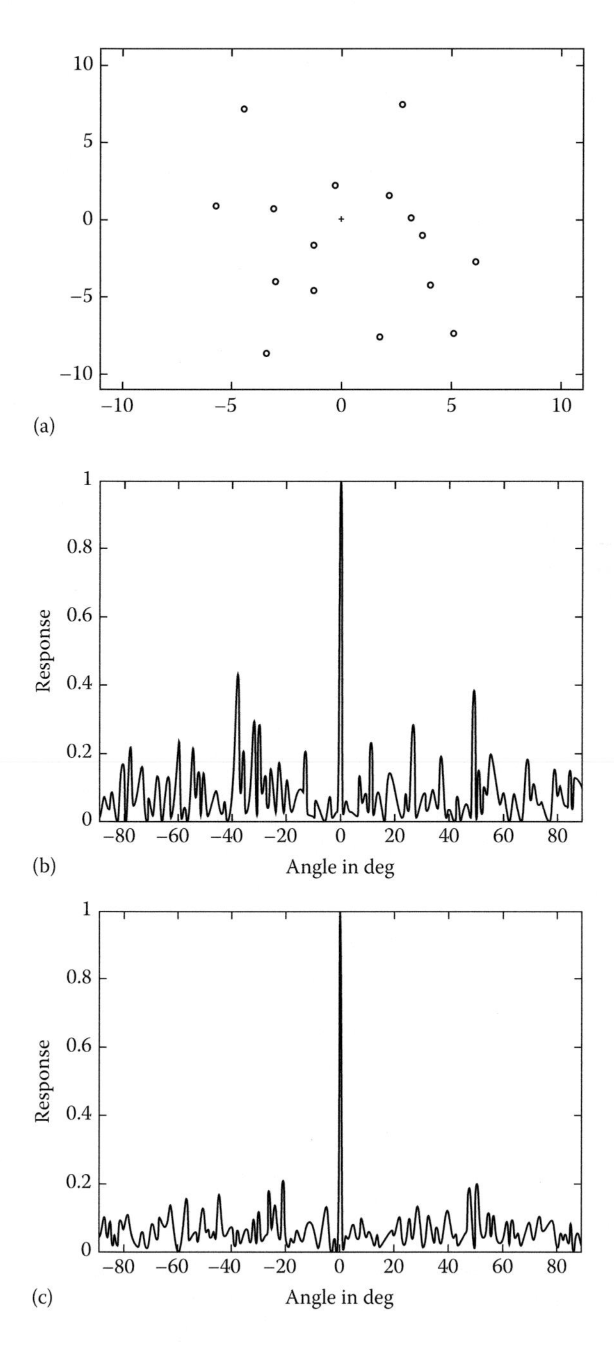

FIGURE 2.10 (a) Sensor cluster around an anchor node; (b) Directional response (normalized); (c) Averaged response over three adjacent wavelengths (0.09, 0.1 and 0.11 meters).

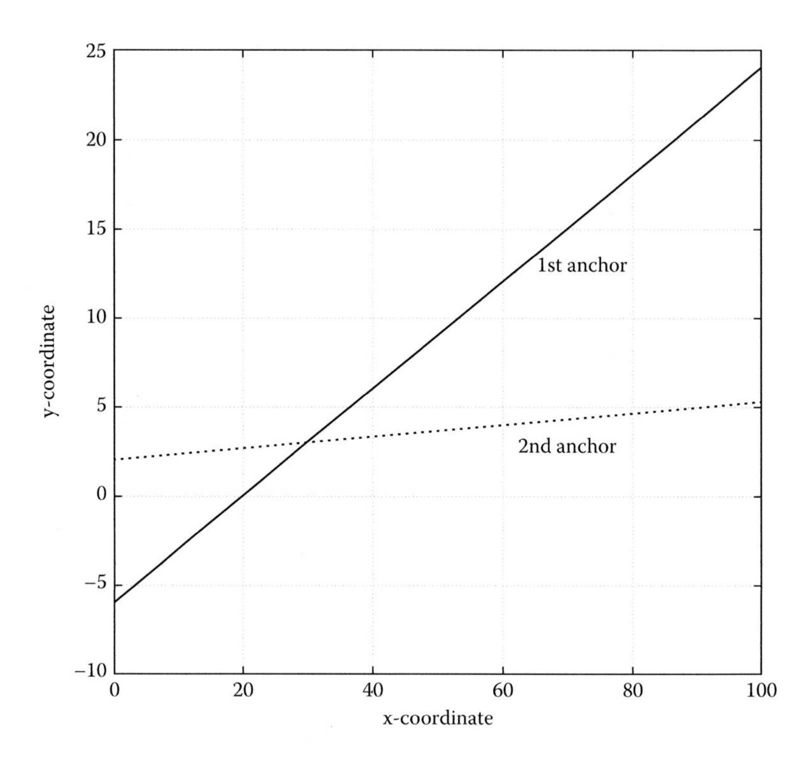

FIGURE 2.11 An intersection of two linear constraints is the location of the transmitter, which is located at (30,3). DoAs were computed using Equation 2.11. The first anchor is at (4,6) and the second anchor is at (6,4). The intersection point correctly shows the actual source position.

2.7 PHASE CHANGE

Consider a monotonic signal of frequency f_1 and wavelength λ_1 ($c = \lambda_1 f_1$). As the signal travels from the transmitter to the receiver separated by a distance d, where $d = \lambda_1 n_1 + r_1$, n_1 is an integer and $r_1 < \lambda_1$ is a fraction of the wavelength, there is a phase change. The phase change ϕ_1 in degrees is related to r_1, but the integer n_1 remains arbitrary

$$r_1 = \lambda_1 \frac{\phi_1}{360}$$

We can express the distance in terms of the phase change and unknown integer n_1

$$d = \lambda_1 n_1 + \lambda_1 \frac{\phi_1}{360} \tag{2.13}$$

Since n_1 is not known, it is not possible to estimate d from Equation 2.13.

This difficulty has been overcome by transmitting another sinusoid of frequency f_2 of wavelength λ_2 [12]. We have one more equation in d:

$$d = \lambda_2 n_2 + \lambda_2 \frac{\phi_2}{360} \tag{2.14}$$

Further details and a numerical example are given in Chapter 3, where three other methods are described to overcome the previously mentioned ambiguity. Among them, the simplest method is to choose f_2, the second frequency, such that

$$n_2 - n_1 = 1$$

$$d < d_{\max}$$

where $d_{max} > d$ is the upper bound on the unknown distance between the transmitter and receiver. It is shown in Chapter 3 that

$$d = \frac{c}{f_2 - f_1}\left(1 + \frac{\phi_2 - \phi_1}{360}\right) \tag{2.15}$$

2.8 INSTANTANEOUS FREQUENCY

Instantaneous frequency is defined as a derivative of phase function of a complex signal; for example, a real signal may be mapped into a complex signal by pairing it with its Hilbert transform as an imaginary part. Instantaneous frequency is different from the conventional frequency defined as the inverse of the period of one oscillation. The instantaneous frequency, in contrast, is defined locally as a derivative of the phase function. This property can be used to detect sudden changes in the frequency. Such a sudden change of frequency propagates throughout the array. The event appears at sensors at different times depending upon the propagation delay. This is the basis for the estimation of ToA or TDoA. We discuss this approach in the next chapter.

2.9 SUMMARY

This final section summarizes the basic idea underlying each section and around which different sections of this chapter are developed. Determination of the distance between a transmitter and sensor is the basic requirement for localization. RSS and ToA are commonly used as distance measures. Though the RSS method is simple and computationally easy to implement, it suffers from channel limitations, such as multiple reflections and attenuation. In the lighthouse effect method, a rotating parallel beam of light illuminates stationary sensors. We measure the duration of time that a sensor remains under illumination. The time duration is inversely proportional to the range. This property has been used to estimate all three coordinates of a sensor independently without having to synchronize the clocks, but it requires LoS visibility.

ToA is a popular measure for distance. It is essential that the transmitter and all sensors are time synchronized. If we can model clock drifts as a linear phenomenon, it is possible to estimate the model parameters and then use the model to correct the observed time instants. If the transmitter is inaccessible, we use TDoA instead, where it is enough to synchronize sensors only. There is yet another possibility; if we have an extra transmitter or a reflector at a known location, the synchronization error can be simply removed by subtraction.

DoA measured using a cluster of sensors around an anchor node within its communication range is linearly related to the transmitter location for a fixed anchor location (Equation 2.12). If we have two such anchors, the transmitter lies at the intersection point. It is generally believed that phase measurements do not yield a unique distance estimate, but with two sinusoids, if the frequencies are properly selected, we can uniquely estimate the distance. Finally, in the last section, we have pointed out that the sudden change in the locally defined frequency, instantaneous frequency, as a derivative of phase function of the corresponding analytic signal, can be used to measure time delays.

REFERENCES

1. X. Li, RSS-based location estimation with unknown pathloss model, *IEEE Wireless Communication*, vol. 5, pp. 3626–3633, 2006.
2. K. Roemer, The lighthouse location system for smart dust, *Proceedings of the 1st International Conference on Mobile Systems, Applications and Services*, pp. 15–30, 2003.
3. M. Hazas and A. Ward, A novel broadband ultrasonic location system, Ubicomp 2002, Gothenburg, Sweden, September 2002.
4. N. Bulusu, J. Heideman, and D. Estrin, GPS-less low cost outdoor localization for very small devices, *IEEE Personal Communications*, vol. 7, pp. 28–34, 2000.
5. P. J. Medelius and S. O. Starr, System and method for locating lightning strikes, US Patent 6,420,862 B2, July 2002.
6. K. E. Bullen, *An Introduction to the Theory of Seismology*, 2nd edition, Cambridge, UK: Cambridge University Press, 1959.
7. M. Broxton, J. Lifton, and J. A. Paradiso, Localization on the pushpin computing sensor network using spectral graph drawing and mesh relaxation, *Mobile Computing and Communications Review*, vol. 10, pp. 1–12, 2005.
8. Y.-C. Wu, Q. Chaudhari, and E. Serpedin, Clock synchronization of wireless sensor networks, *IEEE Signal Processing Magazine*, vol. 28, pp. 124–138, 2011.
9. S. Kim and J.-W. Chong, An efficient TDOA-based localization algorithm without synchronization between base stations, *International Journal of Distributed Sensor Networks*, vol. 2015, pp. 1–5, 2015.
10. P. S. Naidu, *Sensor Array Signal Processing*, 2nd edition. Boca Raton, FL: CRC Press, 2000.
11. D. H. Johnson, and D. E. Dudgeon, *Array Signal Processing: Concepts and Techniques*, Englewood Cliffs, NJ: Prentice-Hall, 1993.
12. S. Assous, C. Hopper, M. Lovell, D. Gunn, P. Jackson, and J. Rees, Short pulse multi-frequency phase-based time delay estimation, *Journal of the Acoustical Society of America*, vol. 127, pp. 309–315, 2010.

3 Estimation of ToA/TDoA/FDoA

Precise estimation of time of delay (time of arrival [ToA]/time difference of arrival [TDoA]) is crucial for localization. In ToA measurements, a known signal is transmitted at a predetermined time. Sensors receive the known signal after some delay depending upon the distance. Precise delay estimation will depend upon clock synchronization at the transmitter and sensors. Also, in digital processing of sampled signals, fractional delay estimation requires interpolation, which is prone to estimation error. In TDoA measurement, the transmitter may be independent; for example, it may be an unknown transmitter, hence no clock synchronization is possible. However, all sensors must be synchronized. In this section, we first describe important methods of time delay estimation, such as generalized cross-correlation, finite impulse response (FIR) filter, and other methods for TDoA estimation. We also describe two methods that do not require synchronization but would, however, require additional features. In the phase-based method, the transmitter will have to transmit two or more tones, and in the frequency difference of arrival (FDoA) method, the transmitter/receiver is in motion.

3.1 GENERALIZED CROSS-CORRELATION

Let $x_m(t)$ and $x_n(t)$ be the discrete outputs of mth and nth sensors. In the absence of any noise, the two outputs are related to each other, $x_n(t)=x_m(t-\tau_0)$, where τ_0 is the relative time delay. The cross-correlation between the two outputs is given by

$$c_{mn}(\tau) = \frac{1}{T-\tau}\sum_{t=0}^{T-\tau} x_m(t)x_n(t+\tau) \qquad T \to \infty \tag{3.1}$$

The cross-correlation function will peak at $\tau=\tau_0$, making it very easy to determine the actual time delay. Ideally, the most desirable signal for delay estimation is the one that produces a perfect spike correlation function. A pseudonoise (PN) signal generated by a block of shift registers possesses such a property. An example of a PN signal generated by five shift registers with one xor gate is shown in Figure 3.1a, and its (auto) correlation function as a function of lag is shown in Figure 3.1b.

A sharp autocorrelation function requires a wideband spectrum. One could use simple white noise as a possible candidate for time delay estimation but, unfortunately, unless the length of signal is very large, the autocorrelation function displays many spurious peaks. PN signals can be used only in active systems, where there is a full control of the kind of signal one can transmit. In a passive system, however,

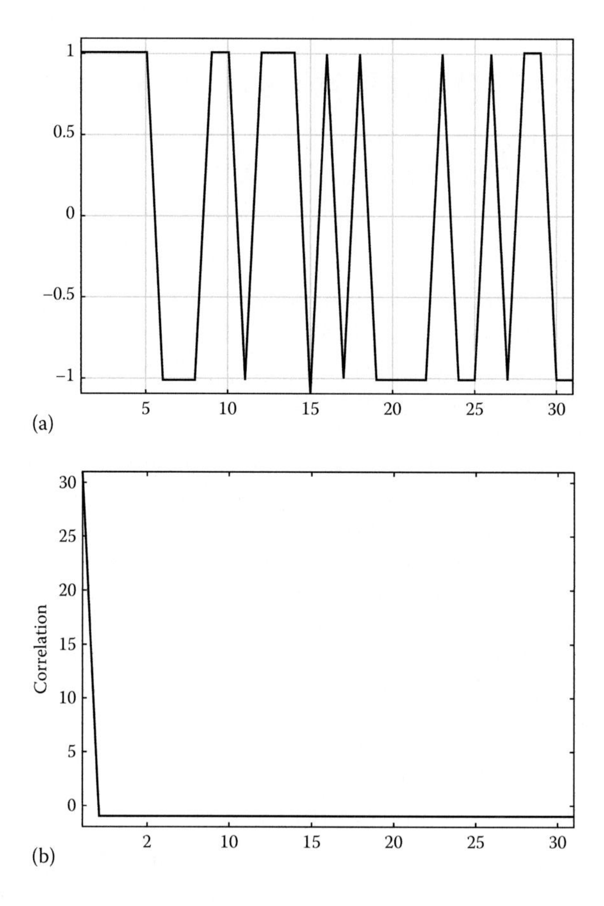

FIGURE 3.1 (a) PN signal of length 31 bits. Five shift registers and one xor gate were used to generate the PN signal. (b) Auto correlation function of PN signal of length 31 bits (shown in Figure 3.1a). It has a peak of height of 31 at zero lag and –1 at all other lags.

this is not possible. The transmitter is quite independent of the sensor array operator. Nevertheless, signal-processing techniques have been designed to essentially flatten the received signal spectrum. In Figure 3.2, a commonly used scheme for time delay measurement is sketched.

Where $h_m(t)$ and $h_n(t)$ are filters applied to whiten the spectrum of the sensor outputs. In the absence of noise, these filters are simply given by

$$H_m(\omega) = \frac{1}{\sqrt{S_m(\omega)}} \tag{3.2a}$$

$$H_n(\omega) = \frac{1}{\sqrt{S_n(\omega)}}$$

where $S_m(\omega)$ and $S_n(\omega)$ are the spectrum of mth sensor output and nth sensor output, respectively. It will be necessary to limit the filtering operation to the frequency band

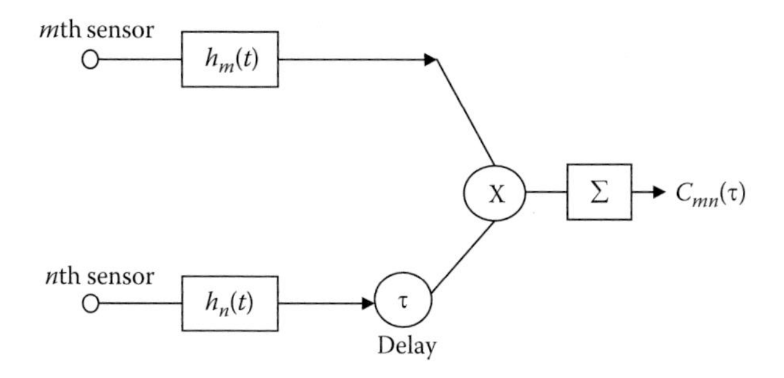

FIGURE 3.2 A commonly used scheme for time delay measurement.

wherein the signal power is significant. Again, in the absence of noise, since the signals differ by delay only, we have $S_m(\omega)=S_n(\omega)$, hence

$$H_m(\omega) = H_n(\omega) = \frac{1}{\sqrt{S_n(\omega)}} \tag{3.2b}$$

This is known as the Roth processor [1]. In the presence of uncorrelated noise in the sensor outputs, the required pre-whitening filters are

$$H_m(\omega) = H_n(\omega) = \frac{1}{\sqrt{|S_{mn}(\omega)|}} \tag{3.2c}$$

where $S_{mn}(\omega)$ is a cross-spectrum between the sensor outputs. Note that, since the noise in the sensor outputs is uncorrelated, the cross-spectrum will be free from the effect of noise, but only in the limiting case of large signal duration. For short signal duration, the presence of even uncorrelated noise can be quite harmful. Along with pre-whitening filters, it is recommended that another filter be used as well, which depends on the signal-to-noise ratio (SNR or snr). Hannan and Thomson [2] have suggested the following pre-whitening filter:

$$\begin{aligned} H_m(\omega) = H_n(\omega) &= \frac{1}{\sqrt{|S_{mn}(\omega)|}} \frac{coh_{mn}^2(\omega)}{1-coh_{mn}^2(\omega)} \qquad coh_{mn}^2(\omega) < 1 \\ &= \frac{1}{\sqrt{|S_{mn}(\omega)|}} snr(\omega) \end{aligned} \tag{3.3}$$

where we have used a relation between coherence and snr [2, p. 126].

$$snr(\omega) = \frac{coh_{mn}^2(\omega)}{1-coh_{mn}^2(\omega)}$$

The pre-whitening filter in Equation 3.3 turns out to be the maximum likelihood (ML) filter under the assumption of Gaussian background noise. The presence of term $snr(\omega)$ ensures that the frequency band where there is strong noise (low snr) is under-weighted. Since both cross-spectrum and coherence are unknown, they may be substituted by estimated quantities, albeit with estimation errors.

The pre-whitening filters given by Equation 3.3 lead to a minimum mean square error in the delay estimation [3]. Estimated delay is unbiased, that is, $E\{(\hat{\tau}-\tau)\}=0$ and its variance is given by

$$E\left\{(\hat{\tau}-\tau)^2\right\}=\frac{1}{T}\frac{1}{\frac{1}{2\pi}\int_{-\infty}^{\infty}\omega^2\frac{snr^2(\omega)}{1+2snr(\omega)}d\omega} \tag{3.4}$$

For a signal with bandwidth B and center frequency ω_c with $\omega_c >> B$ and SNR is constant over the band and >>1, Equation 3.4 reduces to $E\left\{(\hat{\tau}-\tau)^2\right\}\approx \pi/B\ \omega_c^2 T\ SNR$ [20]. It is shown in [1] that the Cramer-Rao bound on the variance of estimated delay is given by right-hand-side of Equation 3.4. Presence of multipaths, particularly those arriving soon after LoS (line of sight) signal, can cause loss of the sharpness of the cross-correlation function, which leads to error in ToA estimation. Another significant limitation arises out of imprecise clock synchronization. We shall elaborate this point in the next chapter.

EXAMPLE 3.1

A simple experiment to compute cross-correlation between the outputs of two microphones and estimate the distance between them from the correlation peak.

The signal source is a simple frequency modulated (FM) radio station. The experimental setup is shown in Figure 3.3. The analog signals were sampled at a rate of 44 kHz using an on-board A/D converter and stored as wave files for later processing. Two thousand samples were extracted from the data files for correlation computation (see Figure 3.4). The Roth processor was implemented, for which the pre-whitening filters, as in Equation 3.2, were used. The resulting correlation function is shown in Figure 3.5. The correlation peak is at 15 units while the theoretical TDoA is 14.33 units (one time unit = 22.72 μsec). The experiment was carried out by a group of final year students led by VVNL Soujanya.

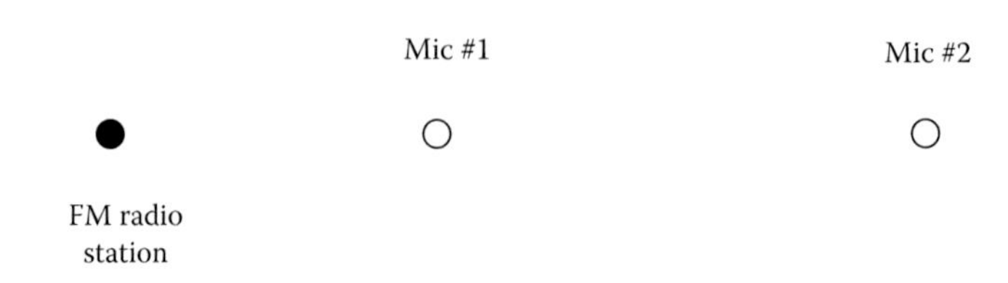

FIGURE 3.3 Experimental setup for delay estimation. The first sensor (microphone) was placed at a distance of 4 ft and the second sensor at 17 ft.

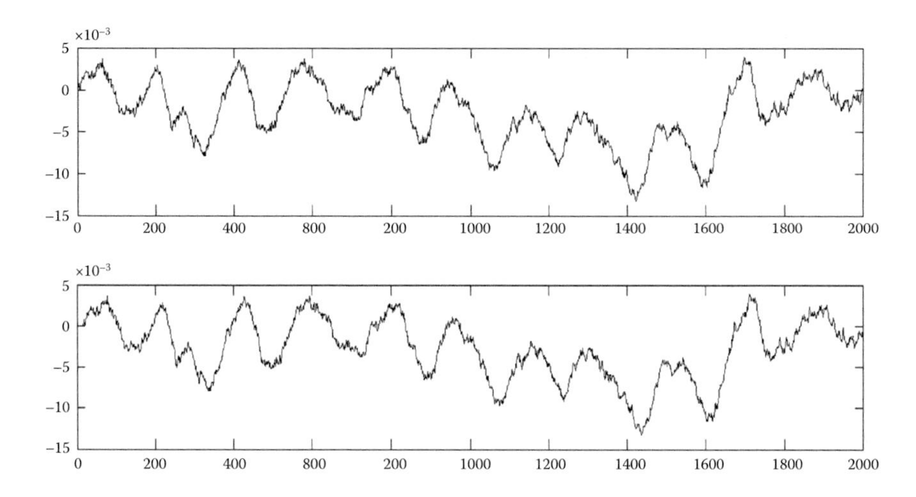

FIGURE 3.4 Sensor outputs. The horizontal axis is time in seconds. The vertical axis is sensor output in arbitrary units.

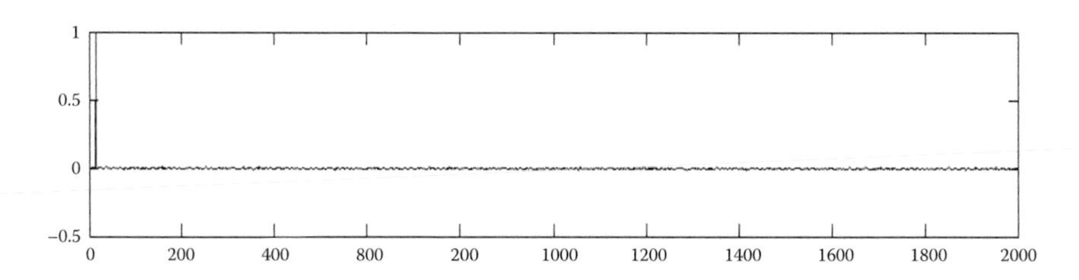

FIGURE 3.5 Roth processor output. The correlation peak stands at a delay of 15 units, while the computed TDoA is 14.33 units.

3.2 TIME DELAY AS FIR FILTER

The delayed signal $x_2(t)$ may be reconstructed from the transmitted signal of the first sensor $x_1(t)$ using a simple interpolation scheme,

$$x_2(t) = x_1(t+(l+f)T) = \sum_{i=-\infty}^{\infty} x_1(iT) \frac{\sin\left(\frac{\pi}{T}(t+(l+f-i)T\right)}{\frac{\pi}{T}(t+(l+f-i)T)} \tag{3.5}$$

where delay, $\tau = (l+f)T$, where l is any integer and f is a fraction. Let $t = kT$ and $m = k - i$. Equation 3.5 after substitution becomes

$$x_2(k) = \sum_{k-m=-\infty}^{\infty} x_1(k-m) \frac{\sin(\pi(m+l+f)}{\pi(m+l+f)} \tag{3.6}$$

where we have assumed the sampling interval, T=1. Using

$$h(m) = \frac{\sin(\pi(m+l+f)}{\pi(m+l+f)} \tag{3.7}$$

in Equation 3.6 we can express it for finite k in a standard convolution sum form

$$x_2(k) = \sum_{m=-\infty}^{\infty} x_1(k-m)h(m) \tag{3.8}$$

The filter coefficients $h(m)$ are samples of $\sin(\pi(t+(l+f)T)/\pi(t+(l+f)T)$, a sinc function with a maximum at $t+(l+f)T=0$. To locate the maximum of $h(m)$, we can use the interpolation formula

$$h(t) = \sum_{m=-\infty}^{\infty} h(m) \frac{\sin\left(\frac{\pi}{T}(t-m)\right)}{\left(\frac{\pi}{T}(t-m)\right)}$$

Let the maximum of $h(t)$ be at t_{max}, then the delay τ is given by

$$\tau = (l+f)T = -t_{max} \tag{3.9}$$

Alternatively, without having to do interpolation, we can obtain the correct location of the maximum. Let the maximum lie between $-l$ and $-l-1$. From Equation 3.7

$$h(-l) = \frac{\sin(\pi f)}{\pi f} \tag{3.10a}$$

$$h(-l-1) = \frac{\sin(\pi(-1+f)}{\pi(-1+f)} = \frac{-\sin \pi f}{\pi(-1+f)} \tag{3.10b}$$

Eliminating $\sin \pi f / \pi f$ from Equation 3.10a and b, we get a solution for f,

$$f = \frac{h(-l-1)}{h(-l-1)+h(-l)} \tag{3.11}$$

The correct position of the maximum is given by Equation 3.8 where we substitute for f from Equation 3.11

$$t_{max} = \left(-l - \frac{h(-l-1)}{h(-l-1)+h(l)}\right)T \tag{3.12}$$

Thus, this method allows estimation of fractional delays from sampled inputs. This was not possible in the GCC method. Now, it remains to show how the filter coefficients can be estimated. For this we need to solve Equation 3.6. For a practical solution, we need to truncate the summation limits from infinity to finite numbers, $\pm P$. Further, we also introduce an extra term representing noise in sensor outputs. We shall give the least-squares solution to the following equation:

$$x_2(k) = \sum_{m=-P}^{P} x_1(k-m)h(m) + \eta(k) \tag{3.13a}$$

We shall first express Equation 3.13a in compact matrix notations. Define the following vectors and matrices:

$$\underset{N \times 1}{\mathbf{X}_2} = \left[x_2(P), x_2(P+1), \cdots x_2(P+N-1)\right]^T$$

$$\underset{2P+1 \times 1}{\mathbf{H}} = \left[h(-P), h(-P+1), \cdots, h(P)\right]^T$$

$$\underset{N \times 2P+1}{\mathbf{X}_1} = \begin{bmatrix} x_1(2P), & x_1(2P-1), \cdots, x_1(0) \\ x_1(2P+1), & x_1(2P), \quad \cdots, x_1(1) \\ \vdots & \vdots \\ x_1(2P+N-1), x_1(2P), & \cdots, x_1(N-1) \end{bmatrix}$$

and

$$\underset{N \times 1}{\boldsymbol{\eta}} = \left[\eta(P), \eta(P+1), \cdots, \eta(P+N-1)\right]^T$$

The compact form of Equation 3.13a is

$$\mathbf{X}_2 = \mathbf{X}_1\mathbf{H} + \boldsymbol{\eta} \tag{3.13b}$$

Minimize the error power with respect to $\mathbf{H}$. The error power is given by

$$\begin{aligned} \boldsymbol{\eta}^T\boldsymbol{\eta} &= (\mathbf{X}_2 - \mathbf{X}_1\mathbf{H})^T(\mathbf{X}_2 - \mathbf{X}_1\mathbf{H}) \\ &= \mathbf{X}_2^T\mathbf{X}_2 - \mathbf{X}_2^T\mathbf{X}_1\mathbf{H} - \mathbf{H}^T\mathbf{X}_1^T\mathbf{X}_2 + \mathbf{H}^T\mathbf{X}_1^T\mathbf{X}_1\mathbf{H} \end{aligned}$$

By differentiating and setting the derivative to zero, we obtain an estimate of the filter coefficients

$$\hat{\mathbf{H}} = (\mathbf{X}_1^T \mathbf{X}_1)^{-1} \mathbf{X}_1^T \mathbf{X}_2 \tag{3.14}$$

The previous estimate turns out to be biased and its variance is given in [4]. As an example, we considered a noisy FM signal

$$x_1(t) = \exp(-j(\omega_0 t + \gamma t^2) + \eta(t)$$

where $\omega_0 = \pi$ and $\gamma = 0.002$. $(t)\,\eta(t)$ is white Gaussian noise (with zero mean and variance equal to 0.1). The output of the second sensor is delayed by 3.3 sec, $x_2(t) = x_1(t-3.3)$. The sensor outputs are sampled at interval T = 1.0 s. Signal duration was selected as N = 1000. The estimates of filter coefficients, given by Equation 3.14, were computed for P = 8. These are plotted in Figure 3.6.

EXAMPLE 3.2

Two microphones were used to receive a speech signal emitted by a loudspeaker. The first microphone was kept close to the speaker and the second microphone was 1.1 m away. The expected delay is 1/3 msec. Received signals are shown in Figure 3.7a,b. The audio signal was sampled at 22 kHz (T = 0.0455 msec). The signal duration used in estimating the filter coefficients was 2000 and the filter length was $2 \times 125 + 1 = 251$. The filter coefficients are shown in the last panel. The peak corresponds to a delay of $73 \times 0.0455 = 0.332$ msec, which is close to the true value, 0.33333 msec. The experiment was repeated for a larger separation in microphones, 3.8 m; but, because of closely spaced multipaths from the walls of the laboratory, the final output was far less dramatic. The experiment was carried out by a group of final-year engineering students led by Vamsi Kiran.

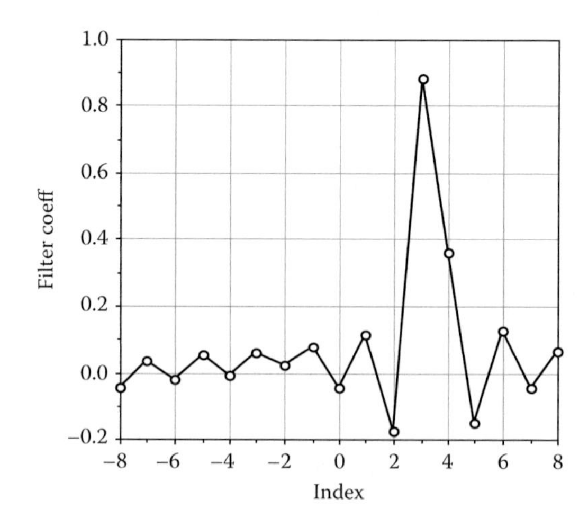

FIGURE 3.6 A plot of estimated filter coefficients. Correct peak lies between index 3 and 4, at 3.28, according to Equation 3.12.

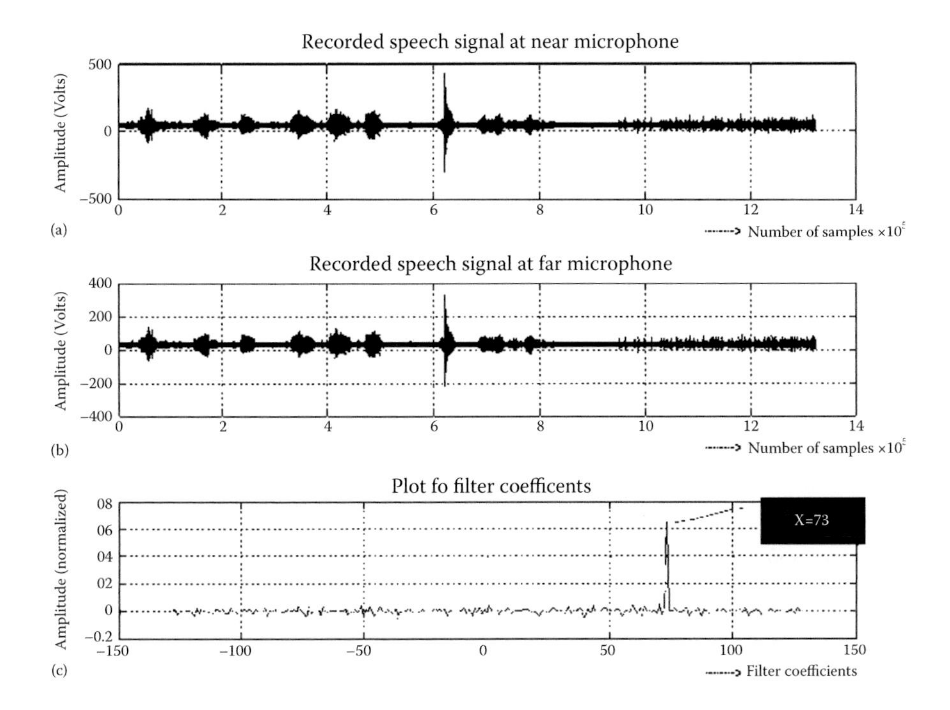

FIGURE 3.7 Speech signal recorded by two microphones set at 1.1 m apart. (a) Output of near microphone; (b) output of far end microphone. The estimated filter coefficients are shown in (c).

3.3 EIGENVECTOR APPROACH

The time delay information is embedded in the spatio-temporal covariance matrix of a DSA. In this section, we show how this information is extracted. There are two possibilities, namely, a near-field case, where a transmitter is close to the array and may even be surrounded by sensors, and a far-field case, where the transmitter is far away from the array (see Figure 3.8). The near-field source is more likely to be strong compared with the far-field source, and hence more easily localized. The sensors are randomly distributed, but their locations are known. In addition, assume that all sensors are time synchronized.

Let M sensors be located at (x_m, y_m), $m=0,1,\ldots,M-1$ in the x–y plane. Arrange all digital outputs in vector form

$$\mathbf{f}(n) = col\left\{f_0(n), f_1(n), \ldots, f_{M-1}(n)\right\} \tag{3.15a}$$

Our goal is, using the array outputs, to localize the transmitter and estimate the signal (i.e., beamforming). Let $s(t)$ be a signal emitted by the transmitter. The output of the mth sensor is given by (assuming line-of-sight [LOS] propagation),

$$f_m(t) = \alpha_m s(t-\tau_m) + \eta_m(t), \ m = 0,1,\ldots,M-1 \tag{3.15b}$$

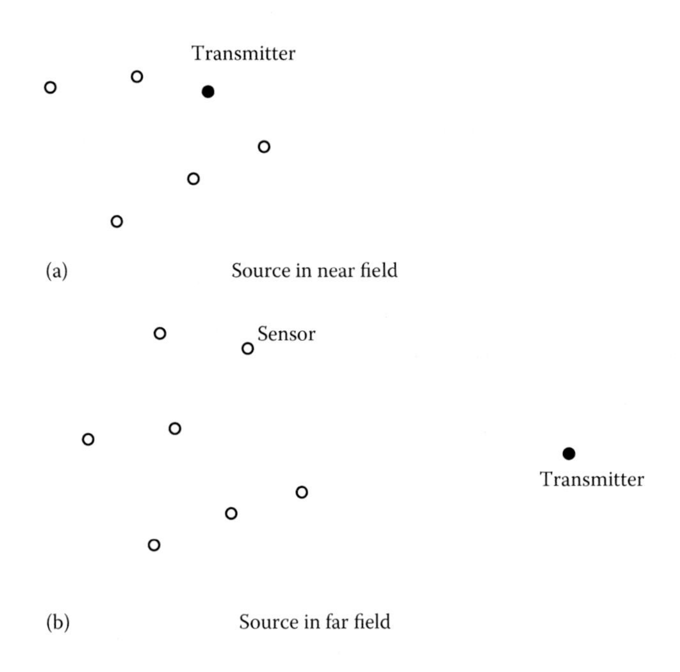

FIGURE 3.8 DSA: (a) near-field case: sensors surround a source. (b) Far-field case: the source is far away from the array, hence the incident wavefront is planar.

where α_m and τ_m are the attenuation and delay, respectively, of the signal and $\eta_m(t)$ is the background noise. Let

$$w_{m,l}, \quad m = 0,1,...,M-1 \text{ and } l = 0,1,...,L-1$$

be weight coefficients. A weighted output array is defined as follows:

$$z(n) = \sum_{m=0}^{M-1} \sum_{l=0}^{L-1} w_{m,l} f_m(n-l) \tag{3.16}$$

which may be expressed in matrix form as

$$z(n) = \mathbf{w}^T \mathbf{f} \tag{3.17}$$

where

$$\underset{ML \times 1}{\mathbf{w}} = \begin{Bmatrix} w_{0,0}, w_{0,1}, \ldots, w_{0,L-1}, w_{1,0}, w_{1,1}, \ldots, w_{1,L-1}, \ldots, \\ \\ w_{M-1,0}, w_{M-1,1}, \ldots, w_{M-1,L-1} \end{Bmatrix}^T$$

$$\mathbf{f} = \begin{Bmatrix} f_0(n), f_0(n-1), \ldots, f_0(n-L+1), f_1(n), f_1(n-1), \ldots, \\ \\ f_1(n-L+1), \ldots, f_{M-1}(n), f_{M-1}(n-1), \ldots, f_{M-1}(n-L+1) \end{Bmatrix}^T$$

$$ML \times 1$$

Note that $\mathbf{w}$ and $\mathbf{f}$ are vectors obtained by *stacking* the weight coefficients and sensor outputs, respectively. The operation of arranging the columns of a matrix, for example, a space–time filter matrix, into a long vector is called stacking. This is often used in array signal processing. The operation of stacking is given by

$$\mathbf{w} \Rightarrow \left\{\mathbf{w}_0^T, \mathbf{w}_1^T, \ldots, \mathbf{w}_{M-1}^T\right\}^T$$

where

$$\mathbf{w}_m = \left\{w_{m,0}, w_{m,1}, \ldots, w_{m,L-1}\right\}^T$$

is the mth column of a space–time filter matrix. We shall call the reverse operation *de-stacking*. Given a long filter vector of size $ML \times 1$, de-stacking involves creating a matrix of size $M \times L$.

The mean output power from Equation 3.16 is given by

$$E\{|z(n)|^2\} = \mathbf{w}^T E\{\mathbf{f}\mathbf{f}^T\}\mathbf{w} = \mathbf{w}^T \mathbf{c}_\mathbf{f} \mathbf{w} \tag{3.18}$$

where $\boldsymbol{c}_f$ is a $ML \times ML$ matrix, known as a multichannel covariance matrix of the array output. It may also be expressed as a block matrix of covariance and cross-covariance matrices. For example, for $M = 3$ (three sensor array) $\boldsymbol{c}_f$ is given by

$$\mathbf{c}_\mathbf{f} = \begin{bmatrix} c_{f_0 f_0} & c_{f_0 f_1} & c_{f_0 f_2} \\ c_{f_1 f_0} & c_{f_1 f_1} & c_{f_1 f_2} \\ c_{f_2 f_0} & c_{f_2 f_1} & c_{f_2 f_2} \end{bmatrix}$$

where

$$\begin{aligned} c_{f_0 f_1} &= E\left\{f_0(t) f_1(t)\right\} \\ &= E\left\{f_0(t) f_0(t + \tau_{01})\right\} \\ &= c_{f_0 f_0}(\tau_{01}) \end{aligned}$$

Note that if there is any clock offset at the transmitter it will not affect $c_{f0f0}(\tau_{01})$.

We like to maximize the mean output power subject to a condition that $\mathbf{w}^T\mathbf{w} = 1$. The solution to the previous maximization problem is straightforward. Solution $\mathbf{w}_{opt}$

is an eigenvector of $\mathbf{c}_f$ corresponding to its maximum eigenvalue [5]. By de-stacking $\mathbf{w}_{opt}$, we obtain the optimum weight matrix, whose rows are the desired filters. The role of filtering is essentially to align all sensor outputs so that they are in phase. By combining all such aligned outputs, we form a beam pointing to the signal source or the strongest source when there is more than one signal source. Interestingly, it is not necessary to estimate the delays; it turns out to be a case of blind beamformation.

The different weight vectors seem to be derived simply by temporally shifting the weight vector of the reference sensor (which is marked as sensor #0). Thus, the relative ToA or TDoA may be obtained from the required temporal shift of the weight vector of a sensor relative to the reference sensor. Further, the weight vector behaves like a band pass filter with its pass band centered over the spectral peak of the transmitted signal. For a narrow band signal, the TDoA is easily estimated from the phase delay measured at peak frequency relative to that of the weight vector at the reference sensor. This approach will work if the TDoAs satisfy an inequality: $\left|\omega_{peak} TDoA / \Delta t\right| \leq \pi$ where Δt is the sampling interval.

In Equation 3.18, the covariance matrix is computed using ensemble averaging, which in practice is difficult to implement as we have just one observation. This drawback is overcome by replacing ensemble averaging with temporal averaging. Define a data matrix,

$$\mathbf{F} = \frac{1}{\sqrt{N}} \left[\mathbf{f}(0)\, \mathbf{f}(1)\, \mathbf{f}(2) \quad \cdots \quad \mathbf{f}(N-1)\right]_{ML \times N}$$

where

$$\mathbf{f}(k) = \begin{Bmatrix} f_0(n+k), f_0(n+k-1), \ldots, f_0(n+k-L+1), \\ f_1(n+k), f_1(n+k-1), \ldots, f_1(n+k-L+1), \ldots, \\ f_{M-1}(n+k), f_{M-1}(n+k-1), \ldots, f_{M-1}(n+k-L+1) \end{Bmatrix}^T_{ML \times 1}$$

$$k = 0, 1, \cdots, N-1$$

The time averaged correlation matrix is given in terms of the data matrix as

$$\hat{\mathbf{c}}_f = \mathbf{F}\mathbf{F}^T \tag{3.19}$$

EXAMPLE 3.3

In this example, we shall demonstrate some of the favorable properties of Yao's eigenvector based weight vectors for TDoA estimation and blind beamformation. We assume a four-sensor distributed array. The sensors are located at (0,0), (20,1), (15,20), and (0,15); the x- and y- coordinates are in meters. There are two uncorrelated transmitters, s_1 in the near-field range at (10,10) and s_2 in the far-field range at (25, 25), which serves the purpose of an interference transmitter. The amplitude of the signal from the near-field transmitter (s_1) is five times greater than that of the signal from the far-field transmitter (s_2). The distributed array geometry and the position of signal transmitters are shown in Figure 3.9. Both sources emit narrow-band uncorrelated signals with the same center frequency at 25 Hz, and their bandwidth is equal to 10 Hz. The sampling interval was taken as $\Delta t=0.005$ sec. The wave speed was set to 1500 m/sec (sound speed in sea). The data length was 1024 samples and the filter length was L=64. From the eigenvector corresponding to the largest eigenvalue, we have computed filters to be applied to each sensor output. All filters are of the same form except for a relative shift (see Figure 3.10 for two sensors), which enables us to compute the TDoA, which was obtained from the phase of the maximum DFT coefficient. The estimated and actual TDoAs are shown in Table 3.1. Estimated waveforms as a filter output are shown in Figure 3.11a. We have switched off the interference transmitter in the far field. The waveform estimate, shown in Figure 3.11b, is very close to that in (a) (normalized mse is 0.009). This validates our claim that the filter focuses on the strong source.

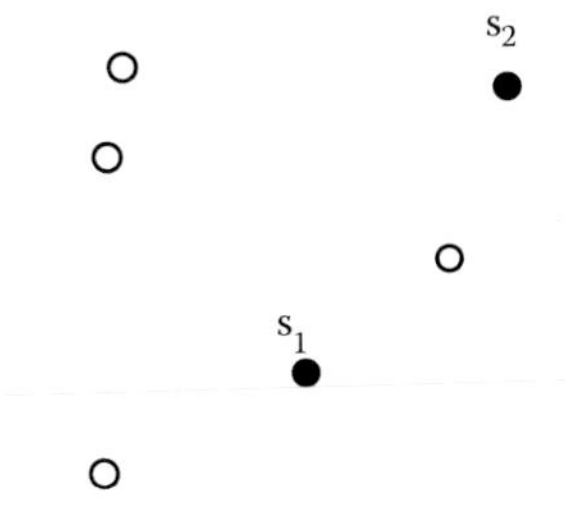

FIGURE 3.9 Four sensor distributed array with two transmitters (filled circles).

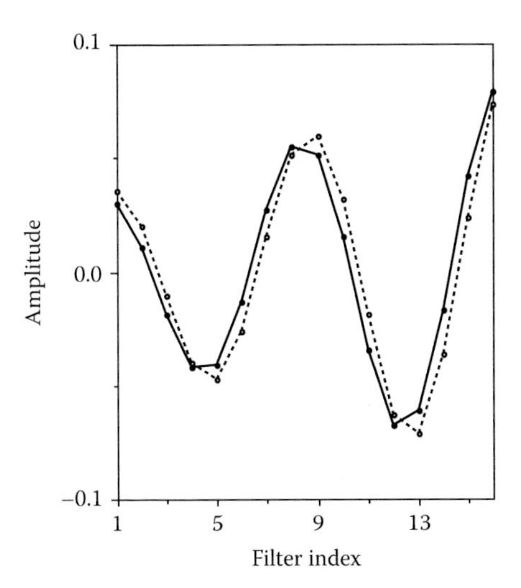

FIGURE 3.10 Filter coefficients for Sensor 1 (thick line) and that for Sensor 3 (dashed line) are shown here.

TABLE 3.1
TDoA Estimate

	Sensor #1	Sensor #2	Sensor #3	Sensor #4
Estimated	0.0	–0.0862	–0.3741	–0.3932
Actual	0.0	–0.0918	–0.3949	–0.3949

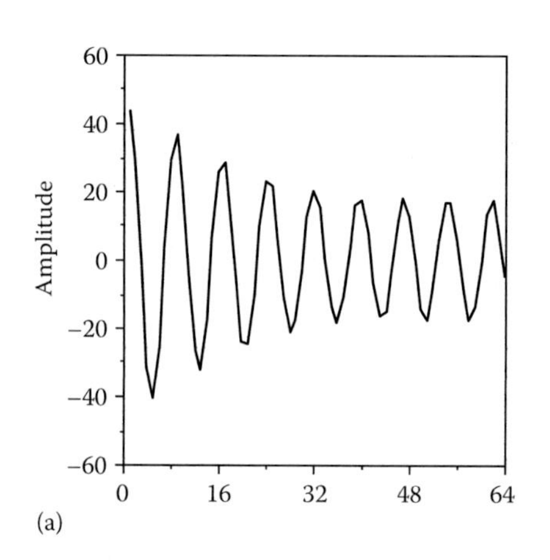

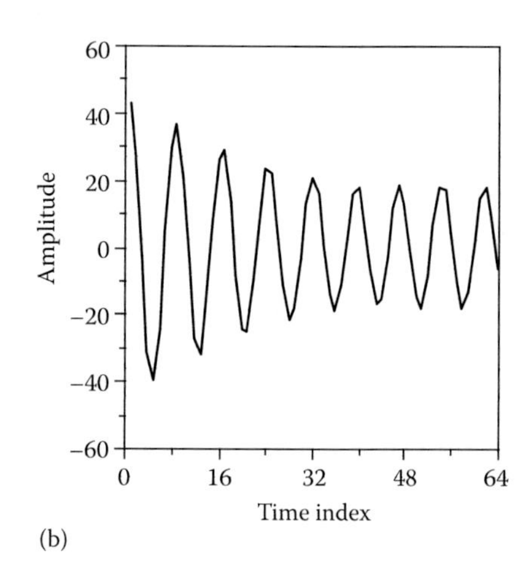

FIGURE 3.11 Estimated waveforms (a) with and (b) without interference.

3.4 SUBSPACE APPROACH

Estimation of multipath delays in a wireless communications channel is an important problem. Separation of overlapping signals is another problem of interest. Bruckstein [6] has recast the problem of overlapping signals into a problem of multiple wavefronts incident on a linear array, and used the idea behind the well-known MUSIC algorithm for the estimation of delays. In the context of a DSA, the sensors receive the same signal with different delays. Treat the output of each sensor as one multipath. Combine all outputs with random weighting into a single output, which would be equivalent to the received signal in a multipath communication channel

$$f(t) = \sum_{m=1}^{M} w_m f_0(t - \tau_m) \tag{3.20}$$

$f_0(t)$ is the known transmitted signal. There are M sensors and τ_m, $m = 1,2,\ldots \boldsymbol{M}$, unknown delays; w_m, $m = 1,2,\ldots \boldsymbol{M}$ are randomly selected weighting coefficients. The weighted signal is now sampled at time instants $t_1, t_2, \ldots t_N$, not necessarily at uniform intervals. We shall express Equation 3.20 in a matrix form. For this let us introduce the following vectors and matrices:

$$\mathbf{f} = \left[f(t_1), f(t_2), \ldots f(t_N)\right]^T$$

$$\mathbf{A} = \begin{bmatrix} f_0(t_1-\tau_1) & f_0(t_1-\tau_2) & \cdots f_0(t_1-\tau_M) \\ f_0(t_2-\tau_1) & f_0(t_2-\tau_2) & \cdots f_0(t_2-\tau_M) \\ \vdots & & \\ f_0(t_N-\tau_1) & f_0(t_N-\tau_2) & \cdots f_0(t_N-\tau_M) \end{bmatrix}$$

and

$$\mathbf{w} = \left[w_1, w_2 \cdots w_M \right]^T$$

Equation 3.20 may be written in terms of the matrices mentioned previously as follows:

$$\mathbf{f} = \mathbf{Aw} + \eta \tag{3.21}$$

Compare Equation 3.21 with 2.18b in [7], which was derived as a signal model for an array of sensors receiving P plane wavefronts. Where the columns of matrix **A** represented the direction vectors, which lie in an array manifold spanned by a direction vector as the direction of arrival (DoA) ranging over $\pm\pi/2$ (for ULA), in the present case, the columns of the **A** matrix are derived by sampling the transmitted signal that has been delayed by an amount equal to the propagation delay to each sensor. The columns of the **A** matrix now lie in a temporal manifold spanned by a sampled signal vector with delay, which covers all possible propagation delays. The temporal manifold is generally known, as the transmitted signal is known, but the rank of **A** is not guaranteed to be equal to M even though the delays are unequal. The rank of **A** depends upon the type of signal. For example, when the signal is constant over the duration of transmission, the rank will be equal to one but, when the signal is rapidly varying, the rank will be equal to M. Therefore, the ability to estimate the delays greatly hinges on the signal structure. Given that the matrix **A** is of full rank, we can readily apply the MUSIC algorithm to estimate the ToA.

The covariance matrix of the data vector **f** is easily computed as

$$E\left\{\mathbf{ff}^T\right\} = \mathbf{A}E\left\{\mathbf{ww}^T\right\}\mathbf{A}^T + E\left\{\eta\eta^T\right\} \tag{3.22}$$

$$\mathbf{C}_f = \sigma_w^2 \mathbf{AIA}^T + \sigma_\eta^2 \mathbf{I}$$

We will assume that the weight vector is drawn from uncorrelated zero mean and unit variance random variables. The observed data vector is thus randomized with a view of making the covariance matrix of full rank in the presence of noise or of rank M in the absence of noise. The noise vector is the sum of noise from each sensor, which is assumed to be uncorrelated. Then, σ_η^2 stands for the sum of noise power from all sensors. The rank of $\mathbf{C}_f - \sigma_\eta^2 \mathbf{I}$ is the same as the rank of $\mathbf{AIA}^T$, which is equal to M. Since N > M (by assumption), there are N–M zero eigenvalues and the corresponding null space shall be orthogonal to the space spanned by the columns of the **A** matrix. Let $\mathbf{v}_{null}$ be a vector in the null space of $\mathbf{C}_f - \sigma_\eta^2 \mathbf{I}$, then,

$$\mathbf{v}_{null}^T \left[f_0(t_1 - \tau) \; f_0(t_2 - \tau) \cdots f_0(t_N - \tau) \right]^T = 0 \quad \tau \in (\tau_1, \tau_2, \cdots, \tau_M)$$

In a MUSIC algorithm, we essentially compute a quantity defined as

$$S_{MUSIC}(\tau) = \frac{1}{\left| \mathbf{v}_{null}^T \mathbf{f}_0(\tau) \right|^2} \tag{3.23}$$

where

$$\mathbf{f}_0(\tau) = \left[f_0(t_1 - \tau) \; f_0(t_2 - \tau) \cdots f_0(t_N - \tau) \right]^T$$

$S_{MUSIC}(\tau)$, known as the MUSIC spectrum, is computed for a large set of delay values. $S_{MUSIC}(\tau)$ will peak wherever $\tau \in (\tau_1, \tau_2, \cdots, \tau_M)$.

EXAMPLE 3.4

A broadband signal, such as a linear FM signal, appears to be an ideal acceptable emitted signal. We consider an FM signal given by

$$f_0(t) = \sin(\gamma\pi(t^2 + t))$$

where γ is a constant (=0.003), which controls the rate of change of frequency. The waveform of the FM signal and its spectrum are displayed in Figure 3.12. We have a three-sensor DSA. The transmitter emits a broadband FM signal. The signals received at Sensors 2 and 3 are delayed by 23 and 25.5 units relative to Sensor 1. We have added unit variance zero mean Gaussian background noise to each received signal. Next, we form a data matrix of 128 samples from three sensors. The size of the data matrix is 128×3. The columns of the matrix are now randomly combined to form a single composite time series. The weighting coefficients used

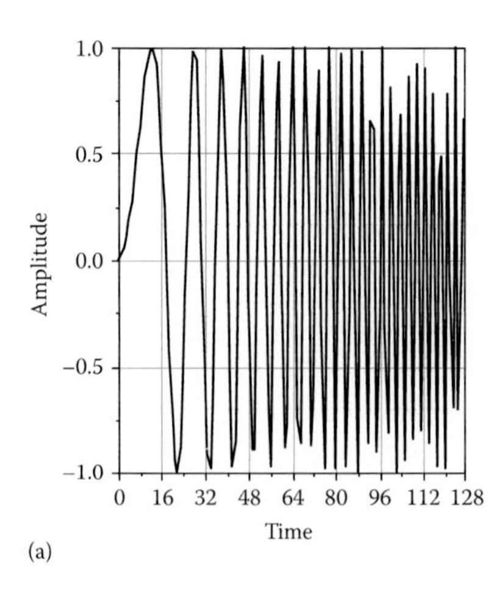

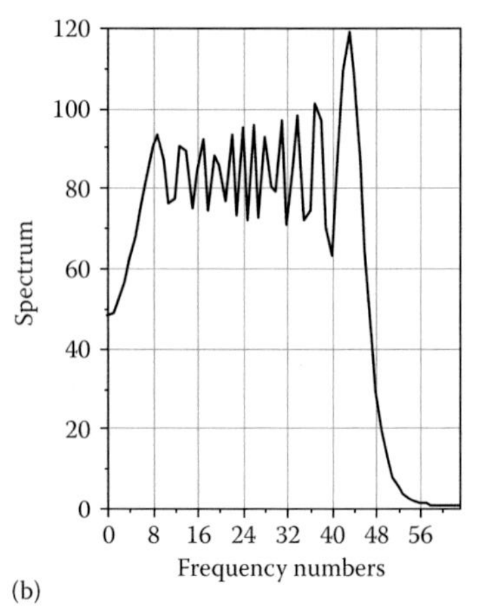

FIGURE 3.12 FM signal and its spectrum.

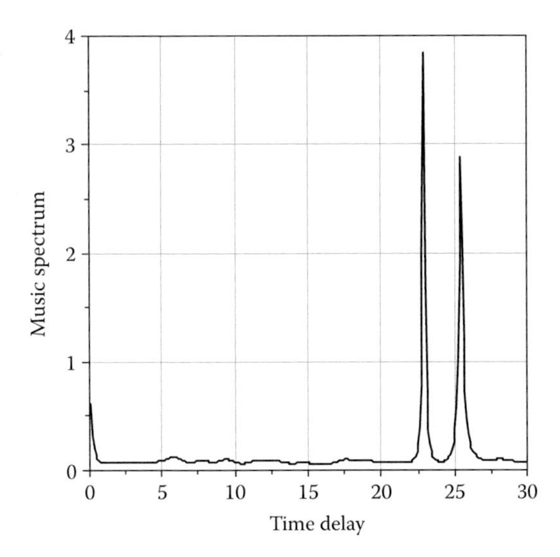

FIGURE 3.13 Music spectrum as a function of time delay. A three-sensor DSA is assumed. The estimated delays at Sensors 2 and 3 are 23.0 (23.0) and 25.5 (25.5). The numbers inside brackets are the actual delays.

are taken from a zero mean unit variance Gaussian random variables. Sixteen such composite time series were formed using independent random weighting coefficients each time. A covariance matrix of size 128×128 is constructed by averaging the outer products of all composite time series. Null space of the covariance matrix is estimated through eigenvalue-eigenvector decomposition. We form a delay vector, analogous to a steering vector, by sampling the transmitted signal (which is known) for a preassigned delay in the range of 0–30 sample units. The MUSIC spectrum was computed according to Equation 3.23. The results are shown in Figure 3.13. The estimated delays are 23.0 and 25.5 in the sample units. The bracketed figures represent actual values.

3.5 CAPON'S MINIMUM VARIANCE

We shall adopt a method of high-resolution detection of ToAs [8] for localization using DSA. In Capon's method, a filter is sought that minimizes the filter power output subject to the constraint that it maintains unit response toward a sensor with prescribed ToA. We shall work in the frequency domain. Equation 3.20 in the frequency domain is given by

$$
\begin{aligned}
F(\omega) &= \sum_{m=1}^{M} w_m F_0(\omega) e^{-j\omega\tau_m} + \sum_{m=1}^{M} w_m \mathrm{N}(\omega) \\
&= \sum_{m=1}^{M} w_m F_0(\omega) e^{-j\omega\tau_m} + \mathrm{N}(\omega)
\end{aligned}
\tag{3.24a}
$$

where F(ω) is the Fourier transform of the randomly combined composite received signals, $F_0(\omega)$ is the Fourier transform of the transmitted signal, and N(ω) is the Fourier transform of the sum of weighted background noise. The weighting coefficients are, as in the last section, selected from zero mean unit variance random variables. Consider a set of discrete frequencies $\omega_1, \omega_1 \ldots, \omega_N$, not necessarily uniformly spaced. We shall now express Equation 3.24a in a matrix form

$$\mathbf{F} = \mathbf{F}_0 \mathbf{A} \mathbf{w} + \mathbf{N} \tag{3.24b}$$

where

$$\mathbf{F} = [F(\omega_1), F(\omega_2), \ldots, F(\omega_N)]^T$$

$$\mathbf{F_0} = diag[F_0(\omega_1), F_0(\omega_2), \ldots, F_0(\omega_N)]$$

$$\mathbf{w} = \left[w_1, w_2, \ldots, w_M\right]^T$$

$$\mathbf{A} = \begin{bmatrix} e^{-j\omega_1\tau_1} & e^{-j\omega_1\tau_2} & \cdots & e^{-j\omega_1\tau_M} \\ e^{-j\omega_2\tau_1} & e^{-j\omega_2\tau_2} & \cdots & e^{-j\omega_2\tau_M} \\ \vdots & & & \vdots \\ e^{-j\omega_N\tau_1} & e^{-j\omega_N\tau_2} \cdots & & e^{-j\omega_N\tau_M} \end{bmatrix}$$

$$\mathbf{N} = [\mathrm{N}(\omega_1), \mathrm{N}(\omega_2), \ldots, \mathrm{N}(\omega_N)]^T$$

Let $\mathbf{h} = \left[h_1, h_2, \ldots, h_N\right]^T$ be a filter vector. The output power is given by

$$\mathbf{h}^H E\left\{\mathbf{F}\mathbf{F}^H\right\}\mathbf{h} = \mathbf{h}^H \mathbf{F}_0 \mathbf{A} E\left\{\mathbf{w}\mathbf{w}^H\right\}\mathbf{A}^H \mathbf{F}_0^H \mathbf{h} + \mathbf{h}^H E\left\{\mathbf{N}\mathbf{N}^H\right\}\mathbf{h}$$

or

$$\mathbf{h}^H \mathbf{C}_f \mathbf{h} = \mathbf{h}^H \mathbf{F}_0 \mathbf{A} \mathbf{C}_w \mathbf{A}^H \mathbf{F}_0{}^H \mathbf{h} + \mathbf{h}^H \mathbf{C}_\eta \mathbf{h}$$

where

$$\mathbf{C}_f = E\left\{\mathbf{F}\mathbf{F}^H\right\}$$

$$\mathbf{C}_w = E\left\{\mathbf{w}\mathbf{w}^H\right\}$$

and

$$\mathbf{C}_\eta = E\left\{\mathbf{N}\mathbf{N}^H\right\}$$

We wish to find the filter **h** that minimizes filter output power

$$\mathbf{h}^H \mathbf{C}_f \mathbf{h} = \min \tag{3.25a}$$

subject to the constraint that the filter maintains a unit response toward a sensor of prescribed delay,

$$\mathbf{h}^H \mathbf{F}_0 \mathbf{a}(\tau) = 1 \tag{3.25b}$$

where $\mathbf{a}(\tau) = \left[e^{-j\omega_1\tau}, e^{-j\omega_2\tau}, \ldots, e^{-j\omega_N\tau}\right]^T$ and τ is the prescribed delay. This is achieved my minimizing the Lagrange function

$$\mathbf{h}^H \mathbf{C}_f \mathbf{h} + \lambda(1 - \mathbf{h}^H \mathbf{F}_0 \mathbf{a}(\tau)) = \min$$

where λ is a Lagrange constant. The solution is given by

$$\mathbf{h} = \frac{\mathbf{C}_f^{-1} \mathbf{F}_0 \mathbf{a}(\tau)}{\mathbf{a}^H(\tau) \mathbf{F}_0^H \mathbf{C}_f^{-1} \mathbf{F}_0 \mathbf{a}(\tau)} \tag{3.26a}$$

and the minimum power is given by

$$CapSpct(\tau) = \frac{1}{\mathbf{a}^H(\tau) \mathbf{F}_0^H \mathbf{C}_f^{-1} \mathbf{F}_0 \mathbf{a}(\tau)} \tag{3.26b}$$

where $CapSpct(\tau)$ is the minimum power output of the filter as a function of the prescribed delay. We shall call this quantity CapSpct in line with the earlier name MuSpct. A plot of the power output as a function of delay shows peaks at true delays. Note that the rank of $\mathbf{C}_f$ is equal to M. Since $N \geq M$, we will find that $\mathbf{C}_f$ is singular and hence its inverse cannot be computed. Diagonally loading the computed covariance matrix is one way to overcome the problem [9].

EXAMPLE 3.5

We illustrate this property through an example. We repeat the experiment described previously. In place of the MUSIC algorithm we have used Capon's minimum variance algorithm. A four-sensor DSA is assumed. The reference sensor, with respect to which all delays are estimated, was scaled down by a factor of 0.01, while the other outputs remained unchanged. The source transmits an FM signal with $\gamma = 0.003$. We have used 128 samples obtained by sampling at unit intervals. The columns of the matrix are now randomly combined to form a single composite time series. The weighting coefficients used are taken from zero mean unit variance Gaussian random variables. Sixteen such composite time series were formed using independently chosen random weighting coefficients each time. Forty-eight discrete frequencies (frequency interval= $\pi/64$) were used in the computation of the $\mathbf{C}_f\mathbf{C}_f$ matrix. The results are shown in Figure 3.14. The

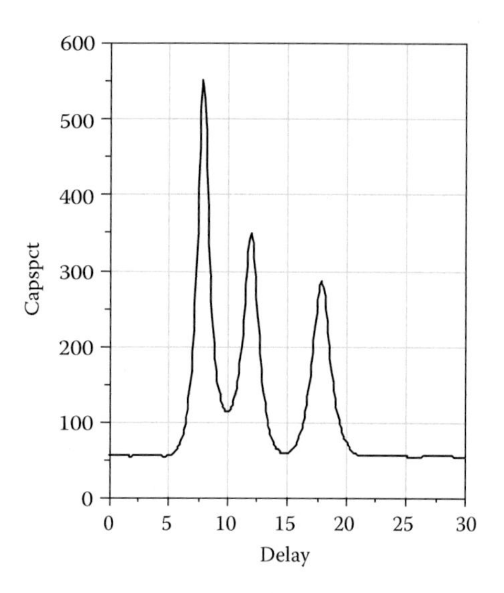

FIGURE 3.14 Capon spectrum of composite signal from four-sensor DSA. The sensor outputs were mixed using weight coefficients drawn from Gaussian random variables. The delays, with respect to the reference sensor, may be estimated from the position of the spectral peaks.

estimated delays are 8.2 (8.0), 12.1 (12.0), and 18.1 (18.0). The numbers within brackets represent to the actual delays.

As for computational complexity, both methods require large matrix (the size is equal to the number of sensors) operations, namely, eigenvector decomposition or matrix inversion. Further, the sensors will have to communicate the raw data to the anchor node, which is likely to be equipped with the required computational power. Thus, high communication overheads and the need for a large computational power seem to be major drawbacks, though the results seem to be very attractive.

3.6 ADAPTIVE NOTCH FILTER

An adaptive notch filter is of interest where the notch frequency is changing with time. A notch filter can be expressed as an infinite impulse response (IIR) filter where the denominator polynomial is a scaled version of the numerator polynomial. Naturally, we can apply the gradient descent method. But this direct form filter does not provide a sharp notch unless the distance to the pole is close to one, in which case it may create filter instability as in any finite precision implementation. Further, since the filter gain depends both on distance to the pole and the angular frequency, the update algorithm may introduce bias in the frequency estimate.

We now consider another approach [10], wherein the influence of the distance to the pole and the angular frequency are decoupled. It is then possible to adapt any one

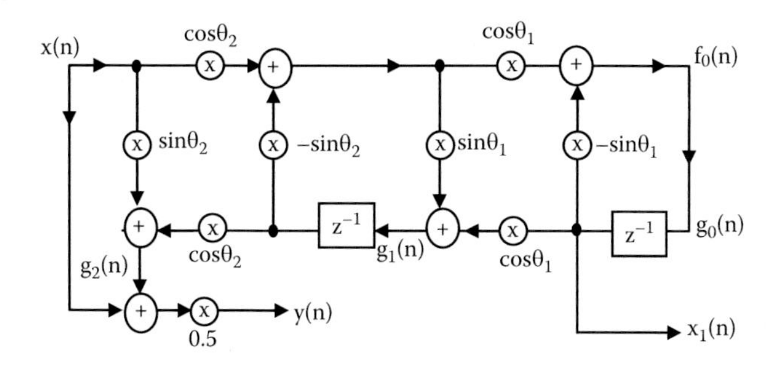

FIGURE 3.15 Planar rotation lattice filter where θ_1 and θ_2 are rotation angles.

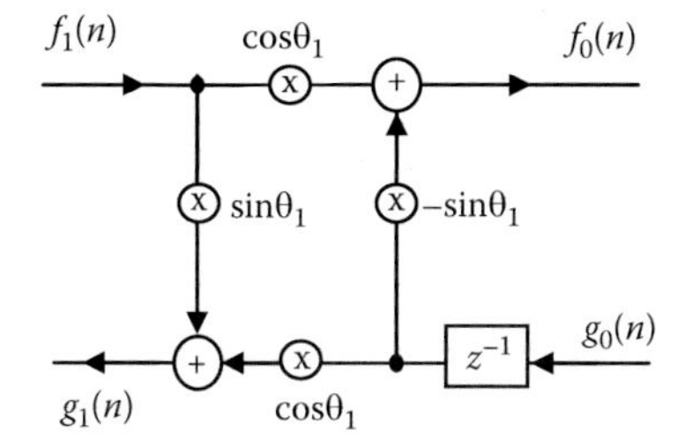

FIGURE 3.16 A single block in lattice filter shown in Figure 3.15.

parameter, for example, the angular frequency, while keeping the distance parameter (distance to pole) fixed. Consider the lattice structure shown in Figure 3.15 comprising planar rotation blocks, where rotation angles θ_1 and θ_2 are adjusted in adaptive realization. The filter structure is stable and numerically well behaved even in time-varying environments [11].

To understand the role of each block in the lattice filter, consider a single block, which is redrawn in Figure 3.16. The output is given by

$$\begin{bmatrix} f_0(n) \\ g_1(n) \end{bmatrix} = \begin{bmatrix} \cos\theta_1 & -\sin\theta_1 \\ \sin\theta_1 & \cos\theta_1 \end{bmatrix} \begin{bmatrix} f_1(n) \\ g_0(n-1) \end{bmatrix}$$

which indeed is equal to a rotation of the input vector by an angle θ_1 in the plane of the input vector, hence the name planar rotation block. Next, we like to evaluate the transfer function (or impulse response function) of the lattice filter. We shall show that the transfer function has a deep notch at a frequency determined by θ_1 and the width of the notch is controlled by θ_2, independently. Let us consider the z-transform of $g_2(n)$ (in Figure 3.15),

$$G_2(z) = X(z)\sin(\theta_2) + z^{-1}G_1(z)\cos(\theta_2)$$

$$G_1(z) = z^{-1}G_0(z)\cos(\theta_1) + (X(z)\cos(\theta_2) - z^{-1}G_1(z)\sin(\theta_2))\sin(\theta_1)$$

$$F_0(z) = (X(z)\cos(\theta_2) - z^{-1}G_1(z)\sin(\theta_2))\cos(\theta_1) - z^{-1}G_0(z)\sin(\theta_1) \tag{3.27}$$

$$F_0(z) = G_0(z)$$

where $G_0(z)$, $G_1(z)$, $G_2(z)$, and $F_0(z)$ are z-transforms of $g_0(n)$ $g_1(n)$, $g_2(n)$, and $f_0(n)$, respectively. In the first equation of Equation 3.27, we replace $G_1(z)$ in terms of $X(z)$ and the parameters in the lattice. After some simplification, we obtain

$$H(z) = \frac{G_2(z)}{X(z)} = \frac{\sin(\theta_2) + \sin(\theta_1)(1+\sin(\theta_2))z^{-1} + z^{-2}}{1+\sin(\theta_1)(1+\sin(\theta_2))z^{-1} + \sin(\theta_2)z^{-2}} \tag{3.28}$$

Note that this form is typical of an all-pass filter [12, p. 249]. The remaining part of the lattice filter in Figure 3.15 is used to realize a notch filter whose response is given by

$$H_{notch}(z) = \frac{Y(z)}{X(z)} = 0.5\left[\frac{(1+2\sin(\theta_1)z^{-1} + +z^{-2})(1+\sin(\theta_2))}{1+\sin(\theta_1)(1+\sin(\theta_2))z^{-1} + \sin(\theta_2)z^{-2}}\right] \tag{3.29}$$

$H_{notch}(z)$ has zeros at

$$z_{1,2}^{-1} = -\sin(\theta_1) \pm j\cos(\theta_1) = e^{\pm j(\theta_1 + \frac{\pi}{2})},$$

which are on the unit circle, at angles

$$\omega_0 = \pm\left(\theta_1 + \frac{\pi}{2}\right)$$

The zeros of the denominator, that is, the poles of the filter Equation 3.28 are given by

$$z_{1,2} = \frac{-\sin(\theta_1)(1+\sin(\theta_2)) \pm j\sqrt{4\sin(\theta_2) - \sin^2(\theta_1)(1+\sin(\theta_2))^2}}{2}$$

The magnitude of the pole is

$$|z_{1,2}| = \sqrt{\sin(\theta_2)}$$

which is a function of the second parameter only. The angular positions of the pole are at $\angle z_{1,2}, \pm(\theta_1 + \pi/2)$, independent of θ_2 only in the limiting case of $\theta_2 \to \pi/2$ (i.e., $b \to 0$). Thus, it is important to note that the location of a notch is *not* controlled by the second parameter; however, it does control the width (or sharpness) of the notch. Let b stand for 3 dB attenuation bandwidth of the notch filter, then $\sin(\theta_2)$ is given by [13],

$$\sin(\theta_2) = \frac{1 - \tan\frac{b}{2}}{1 + \tan\frac{b}{2}} \tag{3.30}$$

Since the location of notch (frequency) and the bandwidth are decoupled (only in the limiting case of $\theta_2 \to \pi/2$), the filter response function for different notch frequencies, but for a fixed bandwidth, $b=0.08$ ($\theta_2=1.176$ radians), becomes independent of the notch frequency. This fact is demonstrated in Figure 3.17. The figure shows IIR notch filter response, given by Equation 3.29, for different frequencies: (1) $\omega_0=\pi/4$ (2) $\omega_0=\pi/2$ (3) $\omega_0=2\pi/3$. The 3 dB widths of all three peaks are practically equal.

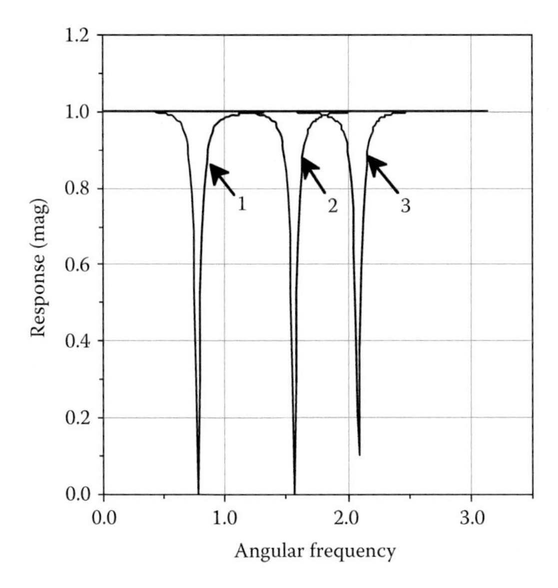

FIGURE 3.17 The figure shows IIR notch filter response, given by Equation 3.29; for different notch frequencies: (a) $\omega_0=\pi/4$, (b) $\omega_0=\pi/2$, and (c) $\omega_0=2\pi/3$. The second parameter is held fixed, $b=0.08$ (bandwidth in radians) or $\theta_2=1.176$.

3.6.1 Adaptation Scheme

In the adaptation scheme, the first parameter θ_1 is estimated using the following updated algorithm [10]:

$$\theta_1(n+1) = \theta_1(n) - \mu(n)y(n)x_1(n) \tag{3.31}$$

where $y(n)$ is filter output and $x_1(n)$ is intermediate output, as shown in Figure 3.15. It is given by

$$X_1(z) = \frac{\cos(\theta_1)\cos(\theta_2)z^{-1}}{1+\sin(\theta_1)(1+\sin(\theta_2))z^{-1}+\sin(\theta_2)z^{-2}}X(z) \tag{3.32}$$

And finally, $\mu(n)$ is step size, which is defined as

$$\mu(n) = \frac{1}{\sum_{k=0}^{n} \lambda^{n-k} x_1^2(n)} \qquad 0 \ll \lambda \leq 1$$

The second parameter, θ_2, which controls the bandwidth, is held fixed.

The product $y(n)x_1(n)$ is not an estimate of the gradient $\partial E\{y^2(n)\}/\partial\theta_1$, and accordingly the update algorithm in Equation 3.31 is not a recursive minimization of an output error cost function. However, it is claimed that the algorithm leads to a stable unbiased parameter estimator in the single sinusoid case and in the multiple sinusoid case, provided the notch bandwidth is sufficiently narrow. The process of adaptation will always converge to the frequency of the strongest sinusoid. The rate of convergence is, however, dependent on the SNR as depicted in Figure 3.18. This property will then enable us to null the dominant sinusoid sequentially.

3.6.2 Delay Estimation

Consider a time variant signal, that is, a signal whose frequency content varies with time. The simplest example is a sinusoidal signal whose frequency changes abruptly at some time instant, say from ω_1 to ω_2. Another common example is a linear (or non-linear) FM signal. Let us now apply the previous adaptive notch filter to track the frequency. The output at the start and at the end will be equal to the frequency of the sinusoid. But as it approaches the transition, output of the tracking filter steadily changes from ω_1 to ω_2. If the time instant of the transition is changed due to propagation delay, the profile of the tracker output will shift laterally by the same amount. This lateral shift then gives us an estimate of the time shift in frequency transition. We first demonstrate the feasibility of this through a simulated case, followed by a physical experiment.

Consider two sensors receiving a time variant signal, where sinusoid frequency changes from $\pi/8$ to $\pi/4$ at time instant 1000 sec. The second sensor receives the same signal with a delay of 100 s relative to the first sensor. It is easy to see that the instants of frequency transition in two sensors will also differ by an amount equal to

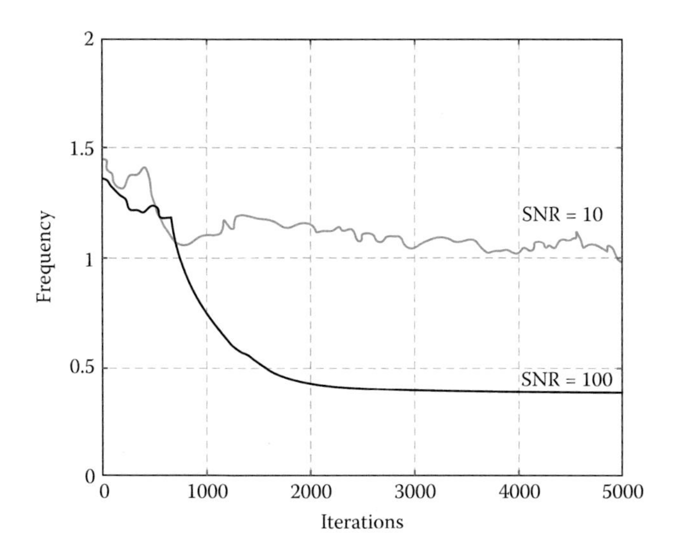

FIGURE 3.18 Convergence of update algorithm in Equation 3.31 for two different SNRs. Single sinusoid of frequency 0.39 (rad/sec) is assumed in the presence of noise. Initial $\mu = 0.001$. The x-axis is in samples or iterations.

TDoA. The signals received at two sensors are shown in the top panel in Figure 3.19. The output of the adaptive filter used to track frequency is shown in the bottom panel (left figure). Notice that the output of the frequency tracker at the second sensor is shifted with respect to that at the first sensor. The shift may be estimated by computing the mean square difference as a function of time shift. The mean square difference is shown in Figure 3.18 on the right bottom panel in Figure 3.19, which shows a minimum at time = 100 sec as expected. A similar exercise was carried out with an FM signal. The delay was correctly estimated from the shift in the frequency tracker profile.

EXAMPLE 3.6

A simple experimental verification of the delay estimation capability of the frequency tracker was arranged. Two microphones were set up 4.5 m apart and the outputs were sampled at 44.1 kHz (sample interval = 0.02267 sec) and stored in a PC. An acoustic source was programmed to generate an FM signal of the same type as shown in Figure 3.19 (bandwidth: 500–2500 Hz). A MATLAB® program was written to compute frequency tracking profile Equation 3.31. In Figure 3.20, the microphone outputs are shown in Figures 3.20a and b, and the tracking profiles are shown in Figure 3.20c. The mean square difference as a function of lateral shift is shown Figure 3.20d. The minimum occurs at a shift of 700 (552) sampling intervals (or iterations), which corresponds to 12.52 ms (that is, $= 552 \times 0.02267$). Theoretical propagation delay is 12.89 ms (speed of sound = 349 m/s at 30°C). The previous experiment was repeated for different microphone intervals. As the experiment was conducted in a laboratory, the noise level was very low, though no attempt to measure the noise power was made. Experimental verification was done by a group of final-year engineering students led by N. Anuradha.

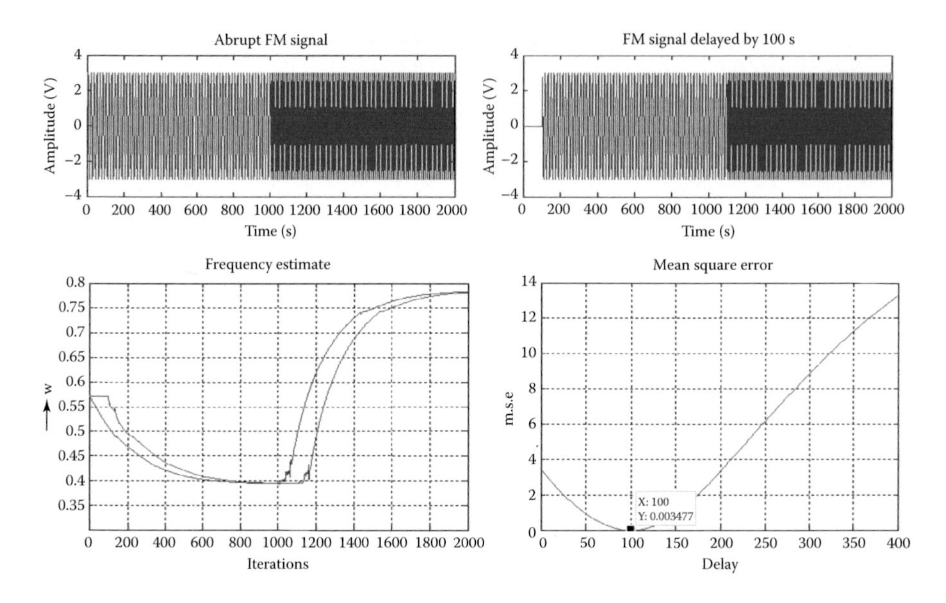

FIGURE 3.19 Signals received at two sensors (top panel). The output of the adaptive filter used to track the frequency is shown in bottom panel (left figure). The mean square difference is shown in left of bottom panel. The location of minimum gives the estimated time shift.

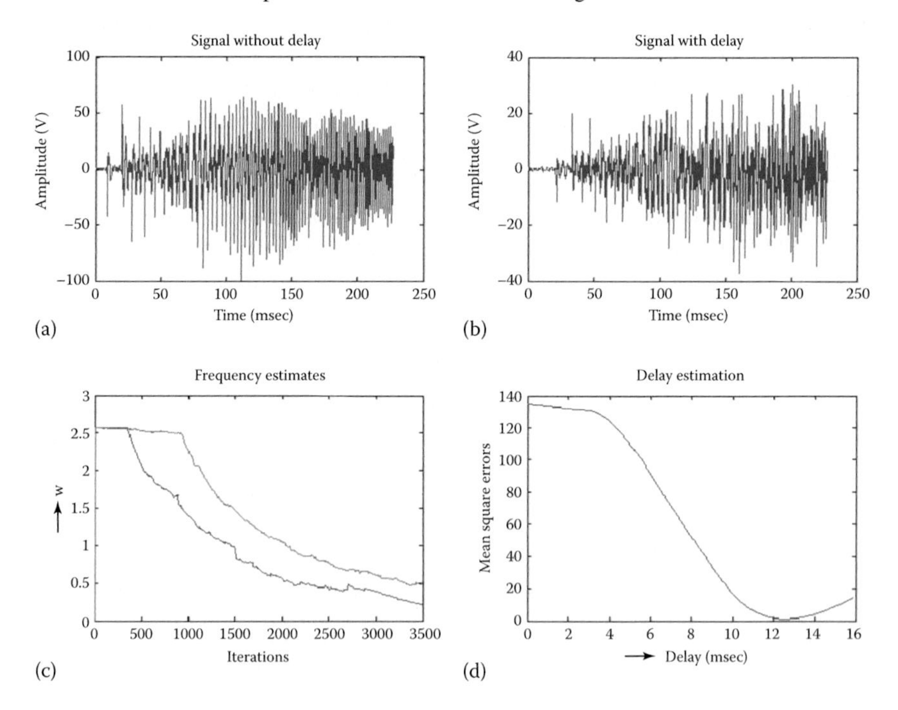

FIGURE 3.20 (a) Near microphone output; (b) far microphone output; (c) frequency tracking profile; and (d) mean square difference as a function of shift (in msec). Estimated delay is 12.52 msec (12.89 msec).

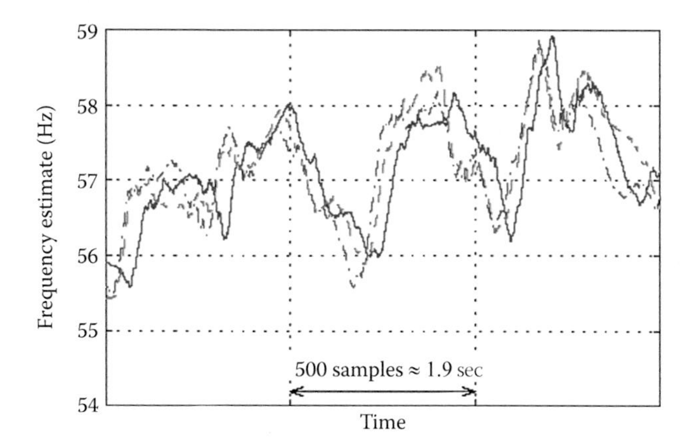

FIGURE 3.21 Frequency tracking profiles at three sensors used for tracking a moving automobile. (From Boettcher, P. and Shaw, G. A., *Proc. of Conf. on Information Fusion*, 1, 2001.)

An experiment to track a moving target (military vehicle) with the help of DSA is described in [14]. Frequency tracking profiles at three sensors are shown in Figure 3.21. As a vehicle runs over a country road, it has to be continuously accelerated and de-accelerated, which results in a changing dominant frequency. The acoustic signal from such a vehicle will have a continuously varying frequency. The frequency tracking profile turns out to be highly undulating and time shifting due to propagation delay (see Figure 3.21). It is quite easy to measure the TDoA from the lateral shift.

3.7 INSTANTANEOUS FREQUENCY SPECTRUM

Closely related to the method described in the last section is instantaneous frequency, defined differently from conventional frequency, as the inverse of the period of a pure tone. The TDoA between two sensors is measured by comparing the instantaneous frequency spectra at two sensors. The transmitter is programmed to transmit a predetermined non-stationary signal; for example, a sinusoidal signal with a sudden jump in the frequency at some known time instant. By observing the instantaneous frequency spectrum, the delay, which depends on the travel time, can be ascertained from relative position of a frequency jump. Instantaneous frequency is defined as a derivative of the phase function of a complex signal; for example, a real signal may be mapped into a complex signal by pairing it with its Hilbert transform as an imaginary part. Such a signal is known as an analytic signal. Instantaneous frequency is different from the conventional frequency defined as the inverse of the period of one oscillation. The conventional frequency is at the center of the Fourier transform, widely used in signal processing. The Fourier transform essentially decomposes a signal into continuous or discrete frequency components. It is ideal for an infinite duration signal, which is an output of a linear system. The frequency and amplitude of every sinusoid remains constant.

The instantaneous frequency, in contrast, is defined locally as a derivative of the phase function. This property can be used to detect sudden changes in the

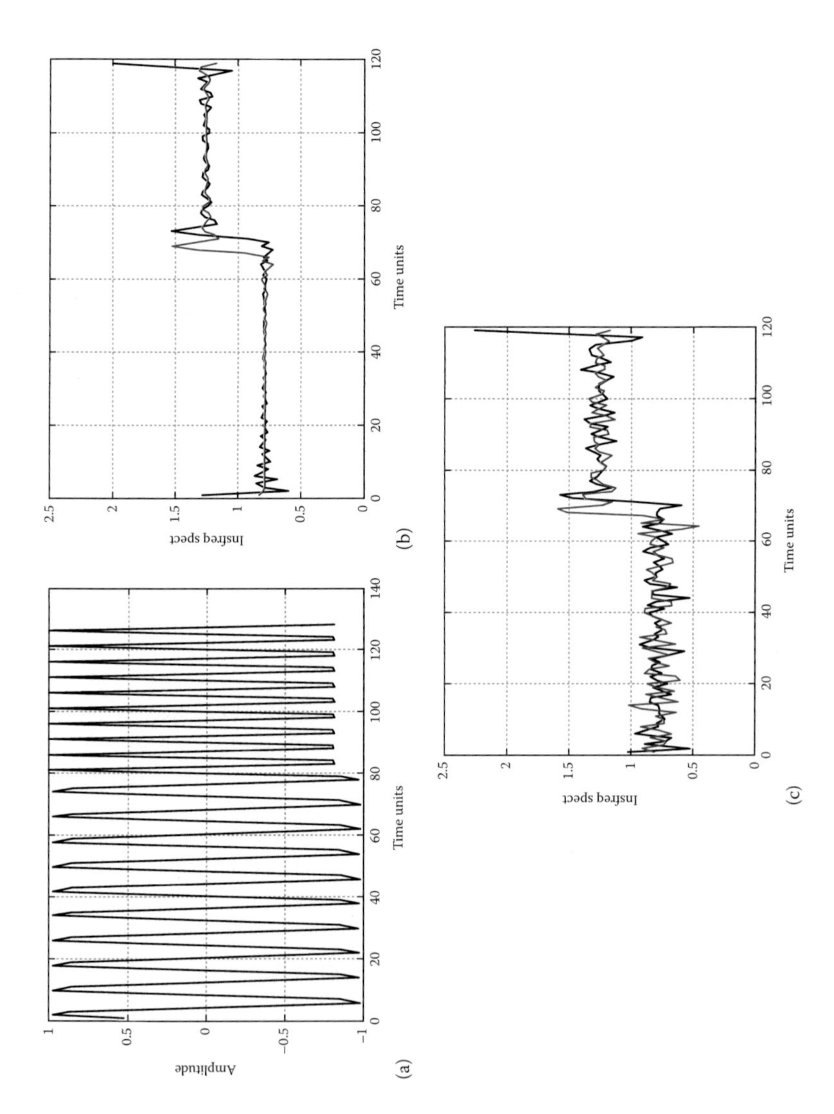

FIGURE 3.22 Two sensors separated by four units of time. (a) Transmitted waveform with abrupt jump in the frequency (0.125–0.16 Hz); (b) instantaneous frequency spectrum (no noise); and (c) instantaneous frequency spectrum with noise (snr = 100).

frequency. Consider a source transmitting a sinusoidal signal where there is a sudden jump in the frequency but all other parameters remain unchanged. Such a signal is received at two sensors; evidently, the ToA at one sensor will be different from that at the second sensor. The delay is best brought out by the instantaneous frequency spectrum of the sensor outputs. There will be a shift, equal to the delay, of the frequency jump in the estimated instantaneous frequency spectrum. This shift can be easily measured, when the noise level is low. We illustrate this possibility through an example.

EXAMPLE 3.7

In Figure 3.22a, the plot of a sinusoidal signal with a sudden jump of frequency from 0.125 Hz to 0.16 Hz (normalized frequency) is shown. The jump is at time units of 65 (64 sec). This is the transmitted signal. It reaches the first sensor after 8 sec and the second sensor after 12 sec. There is a 4 sec TDoA. We compute the analytic signal using the Hilbert transform of each received signal, which becomes the real part of the analytic signal. From the analytic signal, we compute the phase function. A numerical differentiation of the phase function after unwrapping is carried out. We obtain instantaneous frequency (must be positive) as a function of time. We have plotted both instantaneous frequency variations in the same graph (see Figure 3.22b) to emphasize relative shift, which in this case is four time units. The spectrum shown in red refers to the near sensor and that in black refers to the farther sensor. This is a noise-free example. With background noise added, the results deteriorate as shown in Figure 3.22c for snr = 100. For still higher noise levels, say, for snr = 10 the lateral shift is barely discernable.

3.8 PHASE-BASED METHODS

Consider a monotonic signal of frequency f_1 and wavelength λ_1 ($c=\lambda_1 f_1$). The signal travels from a transmitter to a receiver separated by a distance d. Express $d=\lambda_1 n_1 + r_1$ where n_1 is an integer and $r_1 < \lambda_1$ is a fraction of wavelength. Let the phase change in the sinusoidal wave after traveling a distance d and let ϕ in degrees, which is related to r_1,

$$r_1 = \lambda_1 \frac{\phi_1}{360}$$

We can express the distance in terms of phase change

$$d = \lambda_1 n_1 + \lambda_1 \frac{\phi_1}{360} \tag{3.33}$$

We transmit another sinusoid of frequency f_2 of wavelength λ_2. We have one more equation in d,

$$d = \lambda_2 n_2 + \lambda_2 \frac{\phi_2}{360} \tag{3.34}$$

Multiply Equation 3.34 with λ_1/λ_2 and subtract Equation 3.33 from it.

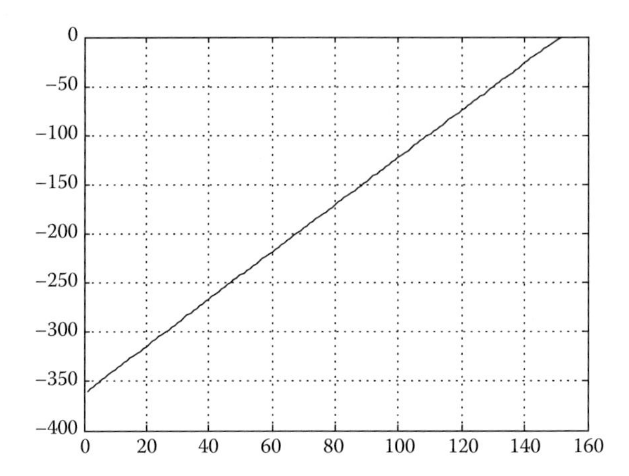

FIGURE 3.23 A plot of $\Delta\phi$ (y-axis) as a function of the distance to target (x-axis). $d_{max} = 1500$ mm and $\Delta n = 1$.

$$\begin{aligned}
\lambda_1/\lambda_2\, d - d &= \lambda_1 n_2 + \lambda_1 \frac{\phi_2}{360} - \lambda_1 n_1 - \lambda_1 \frac{\phi_1}{360} \\
(\lambda_1/\lambda_2 - 1)d &= \lambda_1 \left[(n_2 - n_1) + \frac{(\phi_2 - \phi_1)}{360} \right] \\
d &= \frac{\lambda_1 \lambda_2}{\lambda_1 - \lambda_2} \left[\Delta n + \frac{\Delta\phi}{360} \right] \\
&= \frac{c}{\Delta f} \left[\Delta n + \frac{\Delta\phi}{360} \right]
\end{aligned} \tag{3.35}$$

where $\Delta n = (n_2 - n_1)$, $\Delta\phi = (\phi_2 - \phi_1)$, and $\Delta f = (f_2 - f_1)$. If we restrict the value of Δn to one, we get an unambiguous range, whose upper limit is $d_{max} = c/\Delta f$. Equation 3.35 may be looked upon as a distance traveled by a fictitious sinusoid of wavelength $\Delta\lambda$ ($= d_{max}$), where $\Delta\lambda = c/\Delta f$.

From Equation 3.35 the relative phase of the second sinusoid may be related to the actual distance traveled

$$\frac{d}{d_{max}} = \frac{\Delta\phi}{360} + \Delta n \tag{3.36}$$

The relative phase, $\Delta\phi$, is linear with respect to the distance to the target, as shown in Figure 3.23. Equation 3.36 is independent of frequency, in particular f_1, which is completely a free parameter. Different parameters are obtained as follows:

- First, select the maximum range of interest d_{max}.
- Determine Δf ($= c/d_{max}$) and $\Delta\lambda = c/\Delta f$ ($= d_{max}$).

- Next, select f_1 such that $n_1 = d_{max}/\lambda_1 = f_1 / \Delta f$ is an integer.
- Let $\Delta n = 1$, hence $n_2 = n_1 + 1$.
- Determine $f_2 = f_1 + \Delta f$ and $\lambda_2 = c / f_2$.

That completes all required parameters. For this selection of parameters, the starting phase values are equal to zero, that is, $\phi_1 = \phi_2 = 0$. Indeed, it is necessary that the transmitted sinusoids are all coherent.

EXAMPLE 3.8

We reproduce the example given in [15]. $d_{max} = 1500$ mm, c = 1.5 mm/μsec (in sea water), hence $\Delta f = 1$ kHz. Let $f_1 = 200$ kHz, $\lambda_1 = 7.5$ mm, and $n_1 = 200$. Assume $\Delta n = 1$, hence, $n_2 = 201$, $f_2 = (f_1 + \Delta f) = 201$ kHz, and $\lambda_2 = c/f_2 = 7.462\,\text{mm}$. Note that in this case the initial phase of the sinusoids is zero ($\phi_1 = \phi_2 = 0$). Consider the distance to a target $d = 1000.1234\,\text{mm}$. From Equation 3.36, the expected phase difference is $\Delta\phi = -119.97°$ Consider another example. Let $f_1 = 300$ kHz, and the corresponding wavelength is $\lambda_1 = 5.0$ mm and $n_1 = 300$. Let $\Delta n = 1$, therefore, $n_2 = 301$ and $f_2 = 301$ kHz. Let the distance to a target be 800 mm. The phase change from Equation 3.36 is –168°.

Phase Errors: Phase measurements are always accompanied by errors, largely on account of background noise in the signal. In Equation 3.36, let us express the measured phase difference as a sum of true phase, $\Delta\phi_0$, and measurement phase error, $\delta\phi$,

$$d = \frac{c}{\Delta f}\left[\Delta n + \frac{\Delta\phi_0 + \delta\phi}{360}\right]$$

$$= \frac{c}{\Delta f}\left[\Delta n + \frac{\Delta\phi_0}{360}\right] + \frac{c}{\Delta f}\frac{\delta\phi}{360}$$

$$= d_0 + \delta d$$

where d_0 is the true range and δd is the error in the range estimate. Interestingly, the error is inversely proportional to the frequency interval. To reduce the error in the range estimate, we need to choose a large frequency interval, but this will violate the requirement that $\Delta n \leq 1$ for a unique range estimation. This drawback may be overcome in the second step, where we use the range estimate from the first step to select a better estimate of d_{max}. Using this well-bounded maximum range, we can select a larger frequency interval and consequently a larger Δn $(= d_{max}\Delta f / c)$.

EXAMPLE 3.9

Assume that the phase measurement can be achieved with an accuracy of 0.25°. Let the measured phase, from the previous example (with error), be –120° that is, an error equal to –0.03°. Then the estimated range (from Equation 3.36) is 1000 mm. In the second step, let $\Delta f = 10$ kHz, hence $f_2 = (200 + 10)$ kHz, and $\lambda_2 = c/f_2 = 50/7$ mm. We choose a maximum range as $d_{max} = 1050$ mm; close to the previously estimated coarse range. For the choice of parameters mentioned previously (f_1, f_2 and d_{max}), we obtain $n_1 = 140$ and $n_2 = 147$; hence, $\Delta n = 7$. From Equation 3.36, the computed phase difference is –119.7°. Let the measured phase (with error) be

TABLE 3.2
Simulation Results.

	$f_1-f_0=10$ Hz		$f_2-f_0=100$ Hz	
SNR(dB)	**Mean**	**Var**	**Mean**	**Var**
10	74.942	0.1123	74.99	0.0240
0	74.53	2.0976	75.002	0.0113

Note: T=1 sec, $\Delta t=4\times10^{-4}$, range=75 m, $f_0=1000$, $f_1=1010$ and $f_2=1100$

–119.75°. The estimated range now is (from Equation 3.35) 1000.1042 mm. In the next step, we choose $\Delta f=100$ kHz; hence, $f_2=300$ kHz and $\lambda_2=5$. Let $d_{max}=1005$ mm. For these parameters, $n_1=134$ and $n_2=201$; hence, $\Delta n=67$. The computed phase difference is –117.0384° and let measured phase be –117.0°. Using the measured phase and $\Delta n=67$, the range estimate turns out to be 1000.125 mm. As a matter of fact, we can set Δn to any value in the range $0\leq\Delta n\leq67$, but only the value that yields an estimate closest to that in the previous iteration is admissible. In this case, it turns out to be $\Delta n=67$. In terms of SNR, the range estimates with 10 kHz bandwidth were found to be acceptable even at SNR as low as 0 dB (Table 3.2).

3.8.1 Multitone Signaling

A multitone signal consisting of harmonically related sinusoids is used as a localizing signal:

$$f(t)=\sum_{n=1}^{N}a_n\sin((f_0+n\Delta f)t+\varphi)$$

where f_0 is the fundamental frequency and Δf is the incremental frequency. The delayed version of the previously mentioned signal reaching a sensor at a distance d is

$$f(t)=\sum_{n=0}^{N-1}a_n\sin[(f_0+n\Delta f)(t+\tau)+\varphi]$$

$$=\sum_{n=0}^{N-1}a_n\sin[(f_0+n\Delta f)t+((f_0+n\Delta f)\tau+\varphi))$$

The change of phase due to propagation delay is $(f_0+n\Delta f)\tau$. By selecting $n_1=f_0/\Delta f$ and $n_2=n_1+1$, we can recast multitone ranging into the phase-based ranging with two frequencies. Note that $d_{max}=c/\Delta f$. Consider pairs of frequencies:

$$[(f_0,f_0+\Delta f),(f_0+\Delta f\ f_0+2\Delta f)\cdots(f_0+(N-2)\Delta f\ f_0+(N-1)\Delta f)]$$

Each pair may be used for phase change estimation and range estimation thereof. Thus, we have N range estimates from N frequency pairs. A mean of all these estimates will yield a better estimate of the true range. Here is an example to illustrate the previously mentioned point.

TABLE 3.3
Range Estimates for Different SNRs

SNR dB	d_1	d_2	d_3	d_4	d_5	d_6
0	392.43	414.58	392.54	395.62	411.82	405.39
10	400.66	402.29	402.07	400.49	406.55	397.46
20	401.23	401.83	401,72	403.46	402.80	399.49

Note: Actual range was 402 m. For the lowest SNR the errors in the estimate are within ±2.5%.

EXAMPLE 3.10

We shall consider a signal consisting of seven sinusoids, a carrier frequency, $f_0=30$ MHz, and six harmonics with incremental frequency, $\Delta f=0.3$ MHz. For $c=3.0*10^8$ m/s, the speed of the electromagnetic wave, this will correspond to $d_{max}=1000$ m. All sinusoidal waveforms are assumed to be in phase. The transmitter is located at a distance of 402 m from the receiving antenna. The antenna output is sampled at a rate of 100 Mhz. A block of 6400 samples is used for analysis. For this signal duration, all sinusoids are found to be well resolved in its spectrum. Table 3.3 lists results of range estimates (six trials). The actual range was 402 m. As all trials produced good estimates, it was felt that the estimation of mean and variance might not be necessary. From Table 3.3, we conclude that the phase-based ranging method appears to be quite robust against background noise.

A study was also carried out to evaluate the effect of phase variation in different harmonics. Let ϕ_n be phase in the nth harmonic. We assumed that $\phi_0 = 0$ for the carrier sinusoid. We have modeled $\phi_n = \alpha(r_n - 0.5)2\pi$, where r_n is a uniformly distributed random variable for nth harmonic and α is a constant in the range 0–1.0. In Figure 3.24, we have plotted the average range estimate as a function of coefficient α. When $\alpha=0$, all sinusoids are in phase or coherent. The SNR was set at 10 dB. The average was computed over six pairs of adjacent frequencies. From Figure 3.24, it may be concluded that phase variation must be well within $\pm 0.2\pi$, but, certainly, not outside the range $\pm 0.5\pi$.

3.8.2 Sliding FDFT

Let a transmitter emit a pure tone consisting of multiple wavelengths. Let f_0 be the frequency of the tone and f_s be the sampling frequency. At the receiver, a sliding window of length N is run over the discrete signal and digital Fourier transform (DFT) of finite sequence is computed at every position of the window. Consider fast digital Fourier transform (FDFT) of N past samples starting at time instant n to $N+n-1$

$$X_k(n) = \sum_{l=n}^{N+n-1} x(l)e^{j2\pi k\frac{n-l}{N}} \tag{3.37}$$

where k stands for discrete frequency, which is related to the normal frequency, $f = k f_s / N$.

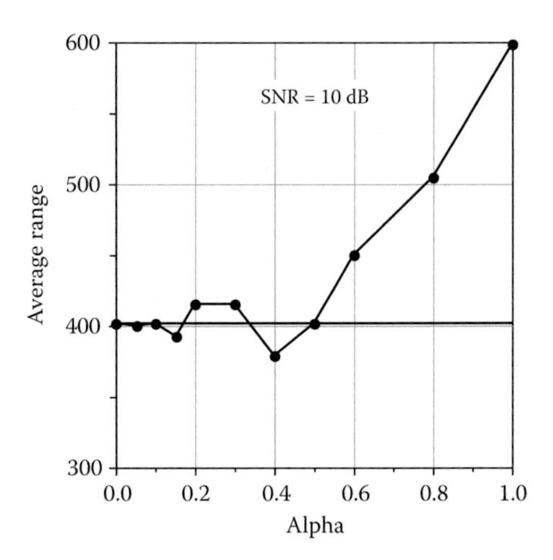

FIGURE 3.24 Average range estimate as a function of coefficient "alpha," which controls the phase in different harmonics. The thick line represents the true range.

We shall now relate sliding FDFT evaluated at two time instants (in units of sampling interval) n and n+1. For this, rearrange terms on the right-hand side of Equation 3.37 as follows [16]:

$$
\begin{aligned}
X_k(n) &= \sum_{l=n}^{N+n-1} x(l) e^{j2\pi k \frac{n-l}{N}} \\
&= x(n) + \sum_{l=n+1}^{N+n-1} x(l) e^{j2\pi k \frac{n-l}{N}} \\
&= x(n) + e^{j\frac{2\pi k}{N}} \left[\begin{array}{l} \displaystyle\sum_{l=n+1}^{N+n-1} x(l) e^{j2\pi k \frac{n-l-1}{N}} + x(N+n) e^{-j\frac{2\pi k}{N}} \\ -x(N+n) e^{-j\frac{2\pi k}{N}} \end{array} \right] \qquad (3.38) \\
&= x(n) + e^{j\frac{2\pi k}{N}} \left[\sum_{l=n+1}^{N+n} x(l) e^{j2\pi k \frac{n-l-1}{N}} \right] - x(N+n) \\
&= x(n) + e^{j\frac{2\pi k}{N}} X_k(n+1) - x(N+n)
\end{aligned}
$$

Equation 3.38 enables us to recursively compute sliding FDFT at the current time instant from the immediate past estimate. For simplicity, let us assume that the signal is periodic, that is, $x(n) = x(N+n)$. Then,

$$X_k(n) = e^{j\frac{2\pi k}{N}} X_k(n+1) \tag{3.39}$$

Similarly, the sliding FDFTs at time instants n and $n+\tau$ are related through

$$X_k(n) = e^{j\frac{2\pi k\tau}{N}} X_k(n+\tau) \tag{3.40}$$

We shall now consider an application of the previous result for delay estimation. Let a transmitter emit a finite duration pure sinusoid of frequency f_s. The duration of transmission is $pT > N$ where T is period ($T = 1/f_0$) and p is an integer. The transmission duration greater than analysis window size N ensures the required symmetry assumed in Equation 3.39. The transmitted signal reaches two sensors with delay τ_1 and τ_2 (in units of sampling interval). Sliding FDFT in Equation 3.38 may be treated as a correlation between the outputs of the first and the second sensor

$$\begin{aligned} X_k(n+\Delta\tau) &= e^{j\frac{2\pi k\Delta\tau}{N}} \left[\sum_{l=n+\Delta\tau}^{N+n+\Delta\tau-1} e^{j2\pi k\frac{l}{N}} e^{j2\pi k\frac{n-l-\Delta\tau}{N}} \right] \\ &= e^{j\frac{2\pi k\Delta\tau}{N}} \left[\begin{array}{l} (N-n+\Delta\tau)e^{j2\pi k\frac{(n-\Delta\tau)}{N}} \\ 0 \le n \le N \end{array} \right] \end{aligned} \tag{3.41}$$

where $\Delta\tau = \tau_2 - \tau_1$ is TDoA. The maximum of the correlation function, $X_k(n)$, lies at $n = \Delta\tau$. Thus the position of the maximum yields the relative delay. The first phase factor will be zero only when TDoA is an integer multiple of T. To see this let us rewrite the phase factor

$$\begin{aligned} e^{j\frac{2\pi k\Delta\tau}{N}} &= e^{j\frac{2\pi f_0\Delta\tau}{f_s}} \\ &= e^{j\frac{2\pi\Delta\tau}{T}} = 1 \quad \text{when} \frac{\Delta\tau}{T} \text{is an integer} \end{aligned}$$

The second factor, which is also complex, has a magnitude equal to $(N+\delta\tau)$ and a phase equal to $2\pi\delta\tau/T$ where $\delta\tau = n - \Delta\tau$, the fractional part of delay. Thus, the phase at peak sliding FDFT of the second sensor output yields the fractional delay, while the position of the peak itself yields the integer part of delay.

EXAMPLE 3.11

Consider a source emitting a sinusoidal signal of frequency 1/8 Hz (normalized) over a duration of 64 sec. We have two sensors separated by TDoA of 16.71 sec. The signal is sampled at a rate of one sample per second. For simplicity, let the first sensor be placed close to the transmitter, so that its output may be treated as the actual transmitted signal, which we require for cross-correlation computation. For the sinusoidal transmitted signal, cross-correlation reduces to sliding FDFT evaluated at a frequency number corresponding to the frequency of the transmitted signal. The

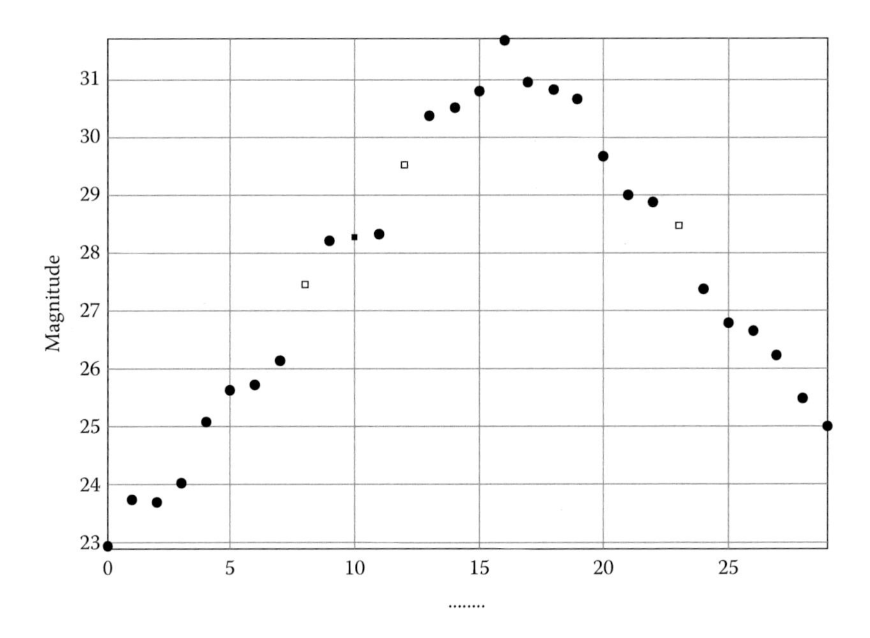

FIGURE 3.25 Magnitude of FDFT at 1/8 Hz for different positions of sliding analysis window. The maximum is at 16 sec.

TABLE 3.4
Effect of Noise on Estimated Fractional Delay

SNR in dBs	Fractional Part (Sec)
17	0.811
20	0.671
30	0.717
∞	0.70999

magnitude of DFT coefficient is plotted as a function of position of the analysis window. The location of the maximum yields the integer part of the delay in units of sample interval (here in seconds). In Figure 3.25, a plot of the magnitude of sliding FDFT (32.00) is shown. The value at the position of the maximum is 0.55763. To get the fractional part of the delay, we use the relation, phase= $2\pi\delta\tau/T$ or fractional delay$=T\,phase/2\pi=0.71$. Thus, the total delay estimated from the sliding FDFT is 16.71 sec, which tallies with the assumed TDoA. The effect of noise is pronounced on the phase estimation. However, the peak position remains practically unchanged. In Table 3.4, we list the estimated fractional delay for different SNRs.

3.9 FREQUENCY DIFFERENCE OF ARRIVAL

Our interest is in estimating the small change in Doppler shift between a pair of sensors. In DSA, sensor-to-sensor communication is limited because of limited available power. Therefore, we shall stress methods that do not require sensor-to-sensor communication. One such possibility is to assume that the mobile transmitter radiates a pure sinusoid or very narrow bandpass signal, and the sensor is equipped with a recursive tone detector. A low-power-consuming FFT processor would be an ideal choice but it is an expensive proposition. A second approach requires a pair of sensors to communicate the received signal to a central processor where we compute time varying delay. FDoA is then derived from the time gradient of time varying delay [17]. This approach is widely used for tracking a moving target in air or underwater. It is important to note that clock synchronization, which is essential for ToA or TDoA, is not required in FDoA measurements. This is indeed a great relief where there is limited available communication bandwidth and power.

3.9.1 Adaptive Frequency Detector

A notch filter is proposed to estimate the frequency of sinusoid in the received signal. Coefficients of the notch filter are adaptively estimated so as to minimize the output power of the notch filter. This is achieved when the notch filter is able to place a zero at unknown frequency. Although the transmitted frequency may be known, the Doppler shift is not known. Hence, the frequency of the sinusoid in the received signal is unknown. An ideal notch filter is an IIR filter with a response function in the z-domain

$$H(z) = \frac{1 - 2\cos(\omega_0)z^{-1} + z^{-2}}{1 - 2r\cos(\omega_0)z^{-1} + r^2 z^{-2}} \tag{3.42}$$

where ω_0 is angular frequency in the range $\pm\pi$ and r a positive constant <1. The previously mentioned transfer function has two zeros at $\pm\omega_0$ on the unit circle and two poles at $\pm\omega_0$ but very close to the zeros. Such a placement of poles and zeros ensures a sharp notch filter [12, p. 245].

An IIR filter shown in Figure 3.26, having a transfer function given by Equation 3.42, may be adopted as a notch filter.

$$H(z) = \frac{1 + k_0(1 + k_1)z^{-1} + k_1 z^{-2}}{1 + a_0(1 + a_1)z^{-1} + a_1 z^{-2}} \tag{3.43}$$

When this is done, the coefficients in Equation 3.43 must satisfy the following equations:

$$\begin{aligned} a_0(1 + a_1) &= rk_0(1 + k_1) \\ a_1 &= r^2 k_1 \end{aligned} \tag{3.44}$$

where $k_0 = -\cos(\omega_0)$ and $k_1 = \leq 1$. We need to estimate first k_0 and k_1 from the input signal so that we can compute a_0 and a_1,

$$a_0 = \frac{rk_0(1 + k_1)}{1 + r^2 k_1} \text{ and } a_1 = r^2 k_1$$

Note that r is a free parameter ($0 < r < 1$), which controls the sharpness of the notch. We adapt k_0 to minimize the mean-square values of p_1 and q_1 in Figure 3.26, and

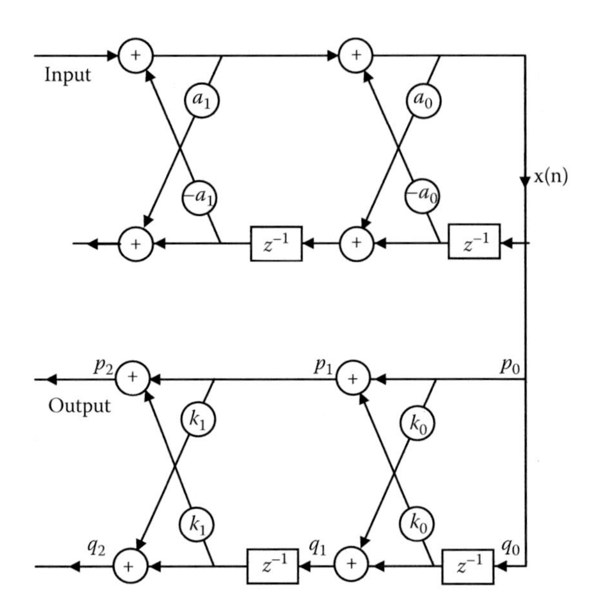

FIGURE 3.26 Adaptive IIR filter structure to implement the transfer function given in Equation 3.43. (Adapted from Cho, C.-M. and Djuric, P. M., *IEEE Trans on Signal Processing*, 42, 3051–3060, 1994.)

similarly k_1 from p_2 and q_2. The following recursive equations [18] are used to adapt k_0 and k_1:

$$k_0(n) = -\frac{C_0(n)}{D_0(n)} \tag{3.45}$$

where $C_0(n)$ and $D_0(n)$ are recursively determined

$$C_0(n) = \lambda C_0(n-1) + p_0(n)q_0(n-1)$$

$$D_0(n) = \lambda D_0(n-1) + 0.5p_0^2(n) + 0.5q_0^2(n-1)$$

and similarly, for k_1

$$k_1(n) = -\frac{C_1(n)}{D_1(n)} \tag{3.46}$$

where

$$C_1(n) = \lambda C_1(n-1) + p_1(n)q_1(n-1)$$

$$D_1(n) = \lambda D_1(n-1) + 0.5p_1^2(n) + 0.5q_1^2(n-1)$$

In the previous equations, λ (<1) is a forgetting factor used to de-emphasize the influence of distant inputs.

The upper half of the IIR lattice filter represents the denominator of the IIR filter. Let $z(n)$ be the input and $x(n)$ the output of this filter section. They are related through a difference equation

$$z(n) = x(n) + a_0(1+a_1)x(n-1) + a_1x(n-2) \tag{3.47}$$

We like to solve this difference equation, where $z(n)$, $n=0,1,\ldots N-1$, is the received signal and $x(n)$, $n=0,1,\ldots N-1$ is the output of the upper circuit in Figure 3.26. The initial conditions are set as $x(-1)=0$ and $x(-2)=0$. We get the following set of linear equations:

$$
\begin{aligned}
x(0) &= z(0) \\
x(1) &= z(1) - a_0(1+a_1)x(0) \\
x(2) &= z(2) - a_0(1+a_1)x(1) - a_1x(0) \\
x(3) &= z(3) - a_0(1+a_1)x(2) - a_1x(1) \\
&\vdots \\
x(N) &= z(N) - a_0(1+a_1)x(N-1) - a_1x(N-2)
\end{aligned}
\tag{3.48}
$$

Because of arbitrary null initial conditions, the early part of the solution will be erroneous, but the errors will soon vanish unless the poles are very close to the unit circle.

The lower section in Figure 3.26 is used to estimate k_0 and k_1

$$
\begin{aligned}
p_0(n) &= q_0(n) = x(n) \\
C_0(n) &= \lambda C_0(n-1) + x(n)x(n-1) \\
D_0(n) &= \lambda D_0(n-1) + 0.5x^2(n) + 0.5x^2(n-1) \\
k_0(n) &= -\frac{C_0(n)}{D_0(n)}
\end{aligned}
\tag{3.49}
$$

and

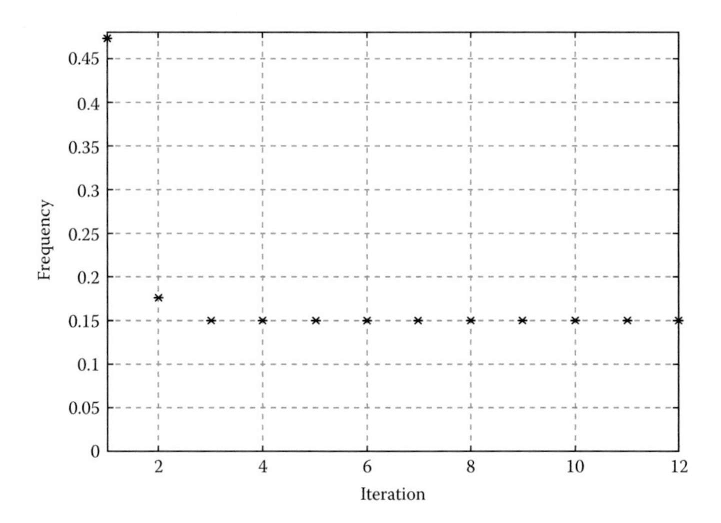

FIGURE 3.27 Convergence of iterative frequency estimate is shown here. Transmitter radiates a pure sinusoid of frequency f_c=0.15 Hz (normalized) with a Doppler shift $f_c/1000 = 0.00015$ Hz. Other parameters are snr=100, sample length=10,000, r=0.99, λ=0.9999, initial values of a_0=1.0 and a_1=0.5. The convergences is rapid, depending upon the snr, as shown in Table 3.5.

$$p_1(n) = p_0(n) + k_0 q_0(n-1) = x(n) + k_0 x(n-1)$$

$$q_1(n) = q_0(n-1) + k_0 p_0(n) = x(n-1) + k_0 x(n)$$

$$C_1(n) = \lambda C_1(n-1) + (x(n) + k_0 x(n-1))(x(n-2) + k_0 x(n-1))$$

$$D_1(n) = \lambda D_1(n-1) + \frac{1}{2}[(x(n) + k_0 x(n-1))^2 + (x(n-2) + k_0 x(n-1))^2]$$

$$k_1(n) = \frac{C_1(n)}{D_1(n)} \tag{3.50}$$

EXAMPLE 3.12

A numerical experiment is carried out to ascertain the convergence of the adaptive algorithm and to verify the accuracy of the estimate. We assumed a Doppler shift of 1000th of transmitted sinusoidal signal. The data length has to be sufficiently large (of the order of the inverse of the Doppler shift). The SNR was varied by varying the power in additive Gaussian noise to the sinusoidal signal. In Figure 3.27, we show the convergence for snr=100. At lower SNR, the convergence is slow

TABLE 3.5
Frequency Estimate Is Quite Accurate Even at as Low an SNR as 1

SNR (linear scale)	Iterations for Convergence	Frequency Estimate
1	9	0.15015
10	4	0.15015
100	3	0.15015

Note: Data length was 10,000 samples; the number of iterations, P, is shown in the second column.

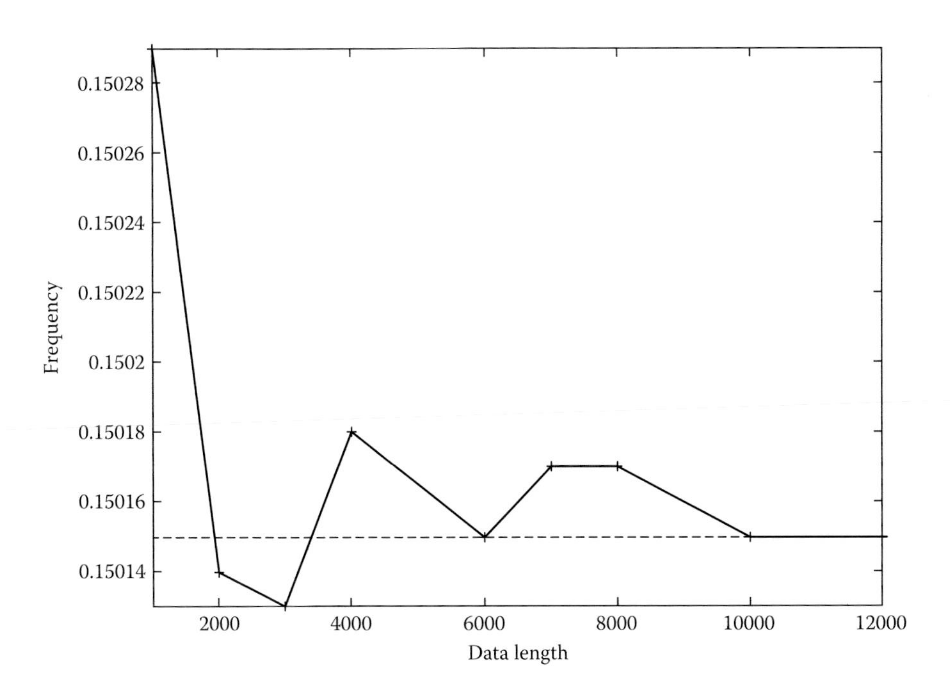

FIGURE 3.28 A plot of frequency to which the iterative process converges as a function of data length. Dashed line represents true value. The snr = 1.0 and other parameters are as in Figure 3.27.

but only marginally, as shown in Table 3.5. Dependence on the initial conditions was briefly studied. We have recomputed the convergence pattern for the same set of parameters as in Figure 3.27 except for a_1, which is now set to 0.8. There is no change in the speed of convergence or in the final result.

It was observed that the length of input data should be of the order of $(1/\delta f)$. For shorter lengths, the iterative process does converge but to a wrong value. A plot of the frequency to which the iterative process actually converges versus data length is shown in Figure 3.28. In all cases, the convergence was achieved after about nine iterations.

3.9.2 Computational Complexity

The algorithm in Equations 3.47 through 3.49 is straightforward, requiring only real additions and multiplications. There are two nested loops, the first one over iterations and the second one over the filtered data from the upper circuit in Figure 3.26. The computational complexity is linear in data length. In fact, it is of the order P*N where P stands for number of iterations and N for data length. Referring to Table 3.5, notice that P varies over 3–9, depending on the SNR. Let us compare this computational complexity with that of the Fourier transform-based approach, which requires computation of Fourier transform of N data points and then location of the position of spectral peak. Fast Fourier transform of N point data length is $N\log_2 N$. For N = 10,000, the computational complexity is of the order of 13 N. For the lattice-based IIR filter approach described here, the computational complexity is of the order of 4 N (for snr = 10). Clearly, the lattice-based IIR filter approach is three times faster than the Fourier transform method.

3.10 SUMMARY

This final section summarizes the very basic idea around which each section is developed.

One of earliest methods for delay estimation is through cross-correlating two signals received at two different sensors. If the signals are white noise, the cross-correlation peak will be sharp and well-positioned. When this is not true, it is possible to whiten the received signal by means of a filter inverse of the square root of the spectrum, which has to be computed from a finite-length received signal; naturally the result will not be ideal. There is another approach to delay estimation; the delayed signal received at the second sensor may be reconstructed from the one received at the first sensor using a simple interpolation scheme. In the case of a band-limited signal, the filter is simply a sinc function with peak at actual delay. In practice, we will have a signal approximately band-limited and of finite length; then we need to resort to a least-squares solution. A practical demonstration of the method using a speech signal from FM radio received at two separated microphones clearly validates this approach.

The time delay information is embedded in the spatio-temporal covariance matrix of a distributed sensor array. This information is extracted from the covariance matrix of a signal obtained by randomly summing over space and time, as in Equation 3.16. The eigenvector corresponding to the largest eigenvalue is the desired filter, whose role is essentially to align all sensor outputs so that they are in phase. By combining all such aligned outputs, we can form a beam pointing to the strongest source, even when there is more than one signal source. Interestingly, it is not necessary to estimate the delays; it turns out to be a case of blind beamformation.

Another closely related approach is to consider DSA as signals from the same source (emitted signal is known) with different delays. Treat the output of each sensor as one multipath (or in discrete form, delay vector). Combine all sensor outputs with random weighting into a single output. Selecting uncorrelated weighting coefficients can generate several such outputs. A covariance matrix is then computed using all outputs. Delay vectors will be orthogonal to the null space of the covariance matrix. The peaks in MUSIC spectrum now correspond to propagation delays. Closely related to the previous is Capon's minimum variance method, where a filter is sought that minimizes the filter power output subject to the constraint that it maintains the unit response toward a sensor with prescribed ToA. The minimum power, Capon spectrum, will have peaks at actual delays, as in the MUSIC spectrum.

We track the dominant frequency in a signal by means of a two-block lattice filter with two parameters. One of them controls the dominant frequency and is adaptively estimated at each time instant. A sudden change in the dominant frequency can be traced to TDoA at a sensor. Closely related to the previously mentioned method is that of estimation of instantaneous frequency, defined differently from conventional frequency, as the inverse of the period of a pure tone. The instantaneous frequency, in contrast, is defined locally as a derivative of the phase function of the analytic signal. As in the adaptive notch filter method, we can determine the time instant of sudden change in the frequency, which can then be used to determine TDoA.

It is generally understood that change in phase of a monotonic signal cannot be used as a measure of distance except within one wavelength. But when we use two selected monotonic signals, change in phase can uniquely measure distance within $c/\Delta f$, where Δf is frequency difference. A multitone signal consisting of harmonically related sinusoids may be used for improved distance estimation by averaging over all adjacent pairs of frequencies.

When either transmitter or receiver is moving, say, at constant speed, there is a small change in Doppler shift between a pair of sensors. FDoA is then derived from the time gradient of time varying delay. Here, time synchronization is not required, which is a great relief.

REFERENCES

1. C. H. Knapp and G. C. Carter, The generalized correlation method for estimation of time delay, *IEEE*, ASSP-24, pp. 320–327, 1976.
2. E. J. Hannan and P. J. Thomson, Estimating echo times, *Technometrics*, vol. 16, pp. 77–84, 1974.
3. M. Azaria and D. Hertz, Time delay estimation by generalized cross correlation methods, *IEEE*, ASSP-32, pp. 280–285, 1980.
4. Y. T. Chan, J. M. Riley, and J. B. Plant, A parameter estimation approach to time-delay estimation and signal detection, *IEEE*, ASSP-28, pp. 8–16, 1980.
5. K. Yao, R. E. Hudson, C. W. Reed, D. Chen, and F. Lorenzelli, Blind beamforming on a randomly distributed sensor array system, *IEEE Journal on Selected Areas in Communications*, vol. 16, pp. 1555–1567, 1998.
6. A. M. Bruckstein, T.-J. Shan, and T. Kailath, The resolution of overlapping echos, *IEEE*, ASSP-33, pp. 1357, 1985.
7. P. S. Naidu: *Sensor Array Signal Processing*, 2nd Edition, Boca Raton, FL: CRC Press, p. 126, 2009.
8. Vidal, M. Najar, and R. Jativa, High resolution time-of-arrival detection for wireless positioning systems, *IEEE, Proc. Vehicular Technology Conference*, vol. 4, pp. 2283–2287, 2002.
9. P. S. Naidu and V. V. Krishna, Improved maximum likelihood spectrum for direction of arrival (DOAs) estimation, *IEEE Conf.*, pp. 2901–2904, 1988.

10. P. A. Regalia, An improved lattice-based adaptive IIR notch filter, *Trans on Signal Processing*, vol. 9, pp. 2124–2128, 1991.
11. A. H. Gray, Jr. and J. D. Market, A normalized digital filter structure, *IEEE Trans. Acoust. Speech, Signal Processing*, vol. ASSP-23, pp. 268–277, 1975.
12. P. S. Naidu, *Modern Digital Signal Processing: An Introduction*, 2nd Edition, New Delhi, India: Narosa, 2006.
13. P. A. Regalia and P. Loubaton, Advances in adaptive orthogonal filtering with applications to source localization, *IEEE* pp. 254–257, 1990.
14. P. Boettcher and G. A. Shaw, A distributed time-difference of arrival algorithm for acoustic bearing estimation, *Proc. of Conf. on Information Fusion*, vol. 1, 2001.
15. S. Assous, C. Hopper and M. Lovell, Short pulse multi-frequency phase-based time delay estimation, *Journal of Acoustical Society of America*, vol. 127, pp. 309–315, 2010.
16. E. Jacobsen and R. Lyons, The sliding DFT, *IEEE Signal Processing Magazine*, pp. 7–80, 2003
17. E. Weinstein and D. Kletter, Delay and doppler estimation by time-space partition of the array data, *IEEE*, ASSP-31, pp. 1523–1535, 1983.
18. J. Makhoul, Stable and efficient lattice methods for linear prediction, *IEEE*, ASSP-25, pp. 423–428, 1977.
19. C.-M. Cho and P. M. Djuric Detection and estimation of DOA's of signals via Bayesian predictive densities, *IEEE Trans on Signal Processing*, vol. 42, pp. 3051–3060, 1994.
20. N. Patwari, J. N. Ash, S. Kyperountas, A. O. Hero III, R. L. Moses, and N. S. Correal, Locating the nodes [cooperative localization in wireless sensor network], *IEEE Signal Processing Magazine*, pp. 54–69, 2005.

4 Localization

Having determined estimates of essential inputs, namely, relative signal strength (RSS), time of arrival (ToA), time difference of arrival (TDoA), direction of arrival (DoA) and frequency difference of arrival (FDoA), we embark on determination of location of transmitter. This is basically a non-linear problem. There are two approaches. In the first approach, the problem is linearized under some permissible assumptions. In the second approach, starting from some initial values, we adaptively update the current values. The adaptive scheme finally converges to the correct estimate, subject to proper selection of the initial values. The selection of the initial values may be based on other available information or determined from other less accurate inputs, such as signal strength or lighthouse effect.

We explore the application of the upcoming concept of compressive sampling (CoSa) [1]. Field strength is measured by an array of a large number of randomly placed and oriented uncalibrated sensors. A point transmitter at an unknown location radiates a random waveform. The sensor output is inverted with a CoSa algorithm as described in [2]. Simple sensor output sign measurements are found adequate for localization provided we have a large number of sensors. The distributed sensor array (DSA) tends to be a collection of a large number of sensors with limited communication power. Therefore, one bit quantization is an advantage in such an environment.

4.1 RECEIVED SIGNAL STRENGTH

The simplest quantity that one can measure with a microphone is signal strength without any reference to neighboring microphones. Consider N microphones (or hydrophones in water) measuring the power of received from an acoustic signal emitted by an acoustic source (transmitter). In a free space, the received signal power is given by

$$s_i(t) = g_i \frac{s_0(t-\tau_i)}{\left|\mathbf{r}_i - \mathbf{r}_0\right|^2} + \varepsilon_i(t) \tag{4.1}$$

where g_i is sensor gain, $S_0(t)$ is signal power emitted by the transmitter, τ_i is delay, and $\mathbf{r}_i$ and $\mathbf{r}_0$ are distance (measured in meters) vectors (column vectors) to the ith sensor and to the transmitter (see Figure 4.1),

$$\mathbf{r}_i = [x_i \ y_i \ z_i]^T$$

$$\mathbf{r}_0 = [x_0 \ y_0 \ z_0]^T$$

$\varepsilon_i(t)$ stands for noise power at the ith sensor. We assume that the sensor gain has been measured before deploying sensors in the field. Thus, g_i is known and for

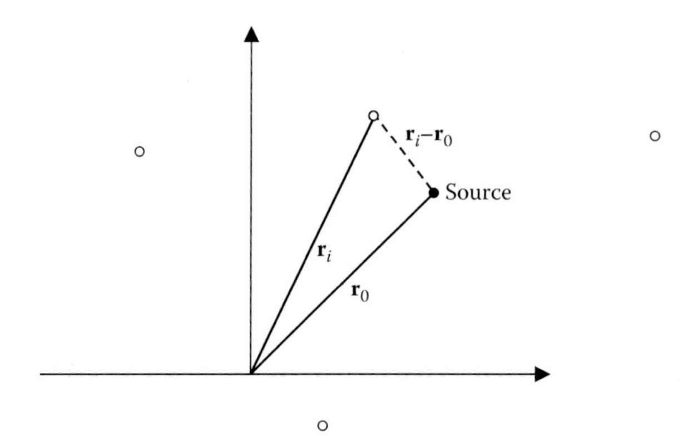

FIGURE 4.1 DSA and acoustic source (transmitter).

simplicity it is set to unity. The power output is then averaged over a time window T, $(T \gg \tau_i)$.

$$\overline{s_i} = \frac{1}{T}\int_T s_i(t)dt = \frac{\overline{s_0}}{\left|\mathbf{r}_i - \mathbf{r}_0\right|^2} + \vec{\varepsilon}_i$$

Next, we compute the ratio of average power at two sensors after subtracting the average noise power, which may be computed when the source is silent.

Let $\overline{\kappa}_{ij}$ stand for the power ratio at the ith and the jth sensor (the reference sensor) at

$$\overline{\kappa}_{ij} = \frac{\overline{s_i} - \vec{\varepsilon}_i}{\overline{s_j} - \vec{\varepsilon}_j} = \frac{\left|\mathbf{r}_j - \mathbf{r}_0\right|^2}{\left|\mathbf{r}_i - \mathbf{r}_0\right|^2} \tag{4.2}$$

Given two sensor locations and the measured power ratio $\overline{\kappa}_{ij}$, the unknown source will lie on a hyper sphere in three dimensions or circle in two dimensions. These are called target location hyper sphere or circle. To show that let us expand Equation 4.2 and rearrange the terms in a form

$$\left|\mathbf{r}_0 - \mathbf{c}_{ij}\right|^2 = \rho_{ij}^2 \tag{4.3}$$

where

$$\mathbf{c}_{ij} = \frac{\mathbf{r}_j - \overline{\kappa}_{ij}\mathbf{r}_i}{1 - \overline{\kappa}_{ij}}$$

and

$$\rho_{ij} = \frac{\sqrt{\overline{\kappa}_{ij}}\left|\mathbf{r}_i - \mathbf{r}_j\right|}{1 - \overline{\kappa}_{ij}}$$

The center of the hyper sphere is located at $\mathbf{c}_{ij}$ and the radius of the hyper sphere is ρ_{ij}. Further, from the definition of $\bar{\kappa}_{ij}$, Equation 4.2, it follows that for a pair of sensors, $\bar{\kappa}_{12} = 1/\bar{\kappa}_{21}$. For three sensors, $\bar{\kappa}_{12}\bar{\kappa}_{23}\bar{\kappa}_{13} = 1$, therefore, given $\bar{\kappa}_{12}$ and $\bar{\kappa}_{13}$, $\bar{\kappa}_{23}$ is redundant. It is shown in [3] that for N sensors there are only N−1 independent power ratios. All the previous results are, however, true only in absence of noise. Three or more circles intersecting at a point uniquely determine the location of a transmitter. To draw three circles, we would need three power ratios or four independent sensors not lying on a straight line. In three-dimensional space, we need four power ratios or five independent sensors not lying on a plane. Here are a few special cases, particularly for two-dimensional sensor distribution. When all sensors fall on a line (collinear), then all centers of the target location circles will also lie on the same line [3]. Circles drawn around these centers depending on the radius may not intersect at all or intersect at two or more points, symmetrically placed about the line joining the sensors. In both cases, transmitters cannot be located. There is however one exception. That is, when the transmitter is in line with the sensors. All target locating circles will become tangential to each other at the transmitter location. Unique transmitter localization is then possible only in this exceptional situation, for any number of sensors (>2).

For $\bar{\kappa}_{ij} = 1$, which, from Equation 4.2 implies $|\mathbf{r}_j - \mathbf{r}_0| = |\mathbf{r}_i - \mathbf{r}_0|$ excluding the trivial case of $i = j$, the hyper sphere becomes a plane equidistant from two sensors. Further simplification of the previous equality leads to

$$\mathbf{r}_0 \cdot \gamma_{ij} = \xi_{ij} \tag{4.4}$$

where

$$\gamma_{ij} = (\mathbf{r}_i - \mathbf{r}_j) \text{ and } \xi_{ij} = \frac{|\mathbf{r}_i|^2 - |\mathbf{r}_j|^2}{2}$$

4.1.1 Transmitter Coordinates

The aim is to estimate the location of a transmitter from the power estimates at multiple sensors distributed arbitrarily over a plane (2D) or in space (3D). Assume that there are N sensors, hence M = N (N−1)/2 sensor pairs. Out of M sensor pairs, let there be M_1 pairs where $\bar{\kappa}_{ij} \neq 1$, therefore, Equation 4.3 applies, and M_2 sensor pairs where $\bar{\kappa}_{ij} = 1$, therefore, Equation 4.4 applies. While Equation 4.4 is linear in source coordinates, Equation 4.3 is non-linear in source coordinates. Non-linear least square estimates have been proposed for estimation of the coordinates [3] but they are computationally intensive, hence they are not quite useful where there is a limited computation power (also battery power). It is possible to linearize Equation 4.3 without any approximation. For this, select a reference sensor and compare the power ratios with respect to the selected reference sensor. Let us consider two sensors and compare their power with respect to that at the reference sensor. Equation 4.3 reduces to

$$
\begin{aligned}
|\mathbf{r}_0|^2 - 2\mathbf{r}_0 \cdot \mathbf{c}_{iref} + |\mathbf{c}_{iref}|^2 &= \rho_{iref}^2 \\
|\mathbf{r}_0|^2 - 2\mathbf{r}_0 \cdot \mathbf{c}_{iref} + |\mathbf{c}_{jref}|^2 &= \rho_{jref}^2
\end{aligned}
\tag{4.5}
$$

where (·) stands for dot product. In Equation 4.5, by subtracting the second equation from the first, we obtain

$$
2(\mathbf{c}_{iref} - \mathbf{c}_{jref}) \cdot \mathbf{r}_0 = \left(|\mathbf{c}_{iref}|^2 - \rho^2_{iref}\right) - \left(|\mathbf{c}_{jref}|^2 - \rho_{jref}^2\right) \tag{4.6}
$$

Considering all possible values of i and j, but excluding $i = j = ref$, we obtain $M = (N-1)(N-2)/2$ equations; for example, for $N = 5$, we get the following set of six equations:

$$
\begin{aligned}
2(\mathbf{c}_{1ref} - \mathbf{c}_{2ref}) \cdot \mathbf{r}_0 &= \left(|\mathbf{c}_{1ref}|^2 - \rho_{1ref}^2\right) - \left(|\mathbf{c}_{2ref}|^2 - \rho_{2ref}^2\right) \\
2(\mathbf{c}_{2ref} - \mathbf{c}_{3ref}) \cdot \mathbf{r}_0 &= \left(|\mathbf{c}_{2ref}|^2 - \rho_{2ref}^2\right) - \left(|\mathbf{c}_{3ref}|^2 - \rho_{3ref}^2\right) \\
2(\mathbf{c}_{3ref} - \mathbf{c}_{4ref}) \cdot \mathbf{r}_0 &= \left(|\mathbf{c}_{3ref}|^2 - \rho_{3ref}^2\right) - \left(|\mathbf{c}_{4ref}|^2 - \rho_{4ref}^2\right) \\
2(\mathbf{c}_{1ref} - \mathbf{c}_{3ref}) \cdot \mathbf{r}_0 &= \left(|\mathbf{c}_{1ref}|^2 - \rho_{1ref}^2\right) - \left(|\mathbf{c}_{3ref}|^2 - \rho_{3ref}^2\right) \\
2(\mathbf{c}_{2ref} - \mathbf{c}_{4ref}) \cdot \mathbf{r}_0 &= \left(|\mathbf{c}_{2ref}|^2 - \rho_{2ref}^2\right) - \left(|\mathbf{c}_{4ref}|^2 - \rho_{4ref}^2\right) \\
2(\mathbf{c}_{1ref} - \mathbf{c}_{4ref}) \cdot \mathbf{r}_0 &= \left(|\mathbf{c}_{1ref}|^2 - \rho_{1ref}^2\right) - \left(|\mathbf{c}_{4ref}|^2 - \rho_{4ref}^2\right)
\end{aligned}
$$

$$
\underset{(6\times3)}{\mathbf{P}} \; \underset{(3\times1)}{\mathbf{r}_0} = \underset{6\times1}{\mathbf{Q}} \tag{4.7}
$$

where

$$
\mathbf{P} = \begin{bmatrix} (\mathbf{c}_{1ref} - \mathbf{c}_{2ref})^T \\ (\mathbf{c}_{2ref} - \mathbf{c}_{3ref})^T \\ \vdots \\ (\mathbf{c}_{1ref} - \mathbf{c}_{4ref})^T \end{bmatrix}
$$

and

$$
\mathbf{Q} = \frac{1}{2} \begin{vmatrix} \left(|\mathbf{c}_{1ref}|^2 - \rho_{1ref}^2\right) - \left(|\mathbf{c}_{2ref}|^2 - \rho_{2ref}^2\right) \\ \left(|\mathbf{c}_{2ref}|^2 - \rho_{2ref}^2\right) - \left(|\mathbf{c}_{3ref}|^2 - \rho_{3ref}^2\right) \\ \vdots \\ \left(|\mathbf{c}_{1ref}|^2 - \rho_{1ref}^2\right) - \left(|\mathbf{c}_{4ref}|^2 - \rho_{4ref}^2\right) \end{vmatrix}
$$

The least-squares solution for $\mathbf{r}_0$ is given by

$$\mathbf{r}_0 = (\mathbf{P}^T\mathbf{P})^{-1}\mathbf{P}^T\mathbf{Q} \tag{4.8}$$

Whenever the power ratio is equal to or close to one, the coefficients in Equation 4.5 become very large, resulting in large numerical error. In this case, Equation 4.3 should be used instead. Note that Equation 4.4 is already a linear equation.

An alternate approach is to treat Equation 4.5 with four independent variables, namely, $[x_0, y_0, z_0, |\mathbf{r}_0|^2]^T$,

$$\mathbf{r}_0 \cdot \mathbf{c}_{iref} - |\mathbf{r}_0|^2 = \frac{|\mathbf{c}_{iref}|^2 - \rho_{iref}^2}{2}, \quad i = 1, 2 \cdots N-1 \tag{4.9}$$

We can express Equation 4.9 in a matrix form

$$\mathbf{P}\theta_0 = \mathbf{Q} \tag{4.10}$$

where

$$\boldsymbol{\theta}_0 = \left[x_0, y_0, z_0, |\mathbf{r}_0|^2 \right]^T$$

and matrices **P** and **Q** are different from those in Equation 4.7). Indeed, they are now,

$$\mathbf{P} = \begin{bmatrix} \mathbf{c}_{1ref}^T & -1 \\ \mathbf{c}_{2ref}^T & -1 \\ \vdots & \\ \mathbf{c}_{N-1ref}^T & -1 \end{bmatrix}$$

and

$$\mathbf{Q} = \frac{1}{2} \begin{bmatrix} |\mathbf{c}_{1ref}|^2 - \rho_{1ref}^2 \\ |\mathbf{c}_{2ref}|^2 - \rho_{2ref}^2 \\ \vdots \\ |\mathbf{c}_{N-1ref}|^2 - \rho_{N-1ref}^2 \end{bmatrix}$$

The least-squares solution of Equation 4.10 is straight forward,

$$\hat{\boldsymbol{\theta}}_0 = (\mathbf{P}^T\mathbf{P})^{-1}\mathbf{P}^T\mathbf{Q} \tag{4.11}$$

4.1.2 Estimation Errors

We look into an estimation of errors in a source location, in particular when we are unable to correctly estimate the noise power. The effect of errors in gain estimation and errors in sensor location, though important, will not be considered here. Let the estimated noise power be $\overline{\varepsilon}_i = \varepsilon_i + \delta\varepsilon_i$ where ε_i is exact noise power and $\delta\varepsilon_i$ is estimation error relative to exact noise power. The error in the power ratio may be related to the error in the estimation of noise power. This is readily derived from Equation 4.2, assuming the error is small

$$\overline{\kappa}_{ij} = \kappa_{ij}(1 - \delta\varepsilon_i + \delta\varepsilon_j) = \kappa_{ij} + \delta\kappa_{ij} \tag{4.12}$$

Assume that $\delta\varepsilon_i$ and $\delta\varepsilon_j$ are zero mean uncorrelated random variables having variances σ_i^2 and σ_j^2 respectively. It follows that δ_{Kij} is also a zero-mean random variable having variance equal to

$$E\{\delta\kappa_{ij}^2\} = \kappa_{ij}^2(\sigma_i^2 + \sigma_j^2)$$

and a cross variance equal to

$$E\{\delta\kappa_{ij}\delta\kappa_{i'j'}\} = 0 \qquad i \neq i' \text{ and } j \neq j'$$

$$E\{\delta\kappa_{ij}\delta\kappa_{ij'}\} = \kappa_{ij}\kappa_{ij'}\sigma_i^2$$

$$E\{\delta\kappa_{ij}\delta\kappa_{i'j}\} = \kappa_{ij}\kappa_{i'j}\sigma_j^2$$

Taking into account a small change in the observed ratio (due to measurement errors), Equation 4.10 may be expressed as

$$\begin{aligned}
(\mathbf{P} + \Delta\mathbf{P})(\boldsymbol{\theta}_0 + \Delta\boldsymbol{\theta}_0) &= \mathbf{Q} + \Delta\mathbf{Q} \\
\mathbf{P}\boldsymbol{\theta}_0 + \mathbf{P}\Delta\boldsymbol{\theta}_0 + \Delta\mathbf{P}\boldsymbol{\theta}_0 &= \mathbf{Q} + \Delta\mathbf{Q} \\
\mathbf{P}\Delta\boldsymbol{\theta}_0 &= \Delta\mathbf{Q} - \Delta\mathbf{P}\boldsymbol{\theta}_0
\end{aligned} \tag{4.13a}$$

Where $\Delta\mathbf{P}$, $\Delta\mathbf{Q}$, and $\Delta\boldsymbol{\theta}_0$ are deviations produced by measurement errors in $\mathbf{P}$, $\mathbf{Q}$, and $\boldsymbol{\theta}_0$ respectively. We have assumed that $\Delta\mathbf{P}$, $\Delta\mathbf{Q}$, and $\Delta\boldsymbol{\theta}$ are small, so that second order and other higher order terms may be neglected. The first order terms are given by

$$\begin{aligned}
\left[\Delta\mathbf{Q}\right]_i &= \frac{1}{2}\left(\left|\mathbf{c}_{i\,ref}\right|^2 - \rho_{i\,ref}^2\right)'\delta\kappa_{iref} \\
\left[\Delta\mathbf{P}\right]_i &= [\mathbf{c}'^{T}_{i\,ref} \;\; 0]\delta\kappa_{iref}
\end{aligned} \tag{4.13b}$$

where (') stands for derivative operation with respect to $\overline{\kappa}_{i\,ref}$ and $\overline{\kappa}_{j\,ref}$. $\delta\kappa_{iref}$ and $\delta\kappa_{iref}$ are defined in Equation 4.12. The required derivatives are easily obtained from the expressions given in Equation 4.3. These are

$$\mathbf{c}'_{i\,ref} = \frac{\mathbf{r}_i - \mathbf{r}_{ref}}{(1-\bar{\kappa}_{iref})^2}$$

$$\left(\left|\mathbf{c}_{iref}\right|^2 - \rho_{iref}^2\right)' = \frac{(\mathbf{r}_i - \mathbf{r}_{ref})^T[(-\mathbf{r}_i + 3\mathbf{r}_{ref})]}{(1-\bar{\kappa}_{iref})^2}$$

Define the following matrices and express Equation 4.13b in a matrix form

$$\mathbf{P}_1 = \begin{bmatrix} \mathbf{c}'^T_{1ref} & 0 \\ \mathbf{c}'^T_{2ref} & 0 \\ \vdots & 0 \\ \mathbf{c}'^T_{N-1ref} & 0 \end{bmatrix}_{N-1\times 4}$$

$$\mathbf{Q}_1 = \frac{1}{2}\left[\left(\left|\mathbf{c}_{i\,ref}\right|^2 - \rho_{i\,ref}^2\right)', i = 1:N-1\right]^T$$

$$\delta\kappa_{ref} = diag\left\{\delta\kappa_{i\,ref},\ i = 1:N-1\right\}$$

Equation 4.13b becomes

$$\Delta\mathbf{P} = \mathbf{P}_1\delta\kappa_{ref}, \quad \Delta\mathbf{Q} = \mathbf{Q}_1\delta\kappa_{ref} \tag{4.14a}$$

Using the previous representation in Equation 4.13a, we obtain

$$\Delta\boldsymbol{\theta}_0 = (\mathbf{P}^T\mathbf{P})^{-1}\mathbf{P}^T(\mathbf{Q}_1 - \mathbf{P}_1\boldsymbol{\theta}_0)\delta\kappa_{ref} \tag{4.14b}$$

Next, we like to compute the mean and variance of $\Delta\boldsymbol{\theta}_0$. Clearly, the mean is zero when $\delta\kappa_{iref}$ is a zero-mean vector. As for the covariance, we obtain the following result:

$$\begin{aligned} \operatorname{cov}\left(\hat{\boldsymbol{\theta}}_0\right) &= (\mathbf{P}^T\mathbf{P})^{-1}\mathbf{P}^T(\mathbf{Q}_1 - \mathbf{P}_1\boldsymbol{\theta}_0) \\ & E\left\{\delta\kappa_{ref}\delta\kappa_{ref}{}^T\right\}(\mathbf{Q}_1 - \mathbf{P}_1\boldsymbol{\theta}_0)^T\mathbf{P}(\mathbf{P}^T\mathbf{P})^{-T} \end{aligned} \tag{4.14c}$$

In the previous approach, we have not exploited the fact that $\left|\mathbf{r}_0\right|^2 = x_0^2 + y_0^2 + z_0^2$. As this constraint is not imposed on the solution, Equation 4.9 is said to be an unconstrained solution. In the absence of noise (i.e., infinite signal-to-noise ratio [SNR]), however, there is no difference. But with finite noise in the measured power ratios, we hope to achieve improved results by imposing the previous constraint.

Note that the estimate and true value relate as

$$\theta(1) = x_0 + \Delta\theta(1)$$

By squaring and rearranging the terms after dropping $|\Delta\theta(1)|^2$ (≈ 0) we obtain

$$\begin{aligned}
&\Delta\theta(1) = \theta(1) - x_0, \quad \text{assuming } |\Delta\theta(1)|^2 \approx 0 \\
&x_0^2 = \theta^2(1) - 2x_0\Delta\theta(1) \\
&y_0^2 = \theta^2(2) - 2y_0\Delta\theta(2) \\
&z_0^2 = \theta^2(3) - 2z_0\Delta\theta(3) \\
&r_0^2 = \theta^2(4) - 2r_0\Delta\theta(4)
\end{aligned} \tag{4.15a}$$

where we have dropped the quadratic terms, $|\Delta\theta(1)|^2$, $|\Delta\theta(2)|^2$, $|\Delta\theta(3)|^2$, and $|\Delta\theta(4)|^2$. After rearranging (Equation 4.15a), we get

$$\begin{aligned}
2x_0\Delta\theta(1) &= \theta^2(1) - x_0^2 \\
2y_0\Delta\theta(2) &= \theta^2(2) - y_0^2 \\
2z_0\Delta\theta(3) &= \theta^2(3) - z_0^2 \\
2r_0\Delta\theta(4) &= \theta^2(4) - r_0^2
\end{aligned} \tag{4.15b}$$

which, in matrix form, is given by

$$\begin{bmatrix} 2x_0 & & & \\ & 2y_0 & & \\ & & 2z_0 & \\ & & & 2|r_0| \end{bmatrix} \begin{bmatrix} \Delta\theta(1) \\ \Delta\theta(2) \\ \Delta\theta(3) \\ \Delta\theta(4) \end{bmatrix} \approx \begin{bmatrix} \theta^2(1) \\ \theta^2(2) \\ \theta^2(3) \\ \theta^2(4) \end{bmatrix} - \begin{bmatrix} 1 & & \\ & 1 & \\ & & 1 \\ 1 & 1 & 1 \end{bmatrix} \begin{bmatrix} x_0^2 \\ y_0^2 \\ z_0^2 \end{bmatrix} \tag{4.16a}$$

$$\mathbf{B}\Delta\boldsymbol{\theta} = \boldsymbol{\theta} - \mathbf{G}\boldsymbol{\phi}_0 \tag{4.16b}$$

where $\boldsymbol{\theta} = [\theta(1)^2\theta(2)^2\theta(3)^2\theta^2(4)]^T$ and $\boldsymbol{\phi}_0 = \begin{bmatrix} x_0^2 & y_0^2 & z_0^2 \end{bmatrix}^T$. The least-squares estimate of $\boldsymbol{\phi}_0$ is given by minimizing $|\Delta\boldsymbol{\theta}|^2$

$$\hat{\boldsymbol{\phi}}_0 = (\mathbf{G}^T\mathbf{B}^{-T}\mathbf{B}^{-1}\mathbf{G})^{-1}\mathbf{G}^T\mathbf{B}^{-T}\mathbf{B}^{-1}\boldsymbol{\theta} \tag{4.17}$$

Note that $E = \{\boldsymbol{\phi}\} = \boldsymbol{\phi}_0$. The error in the estimate, that is, $\delta\boldsymbol{\phi}_0 = \hat{\boldsymbol{\phi}}_0 - \boldsymbol{\phi}_0$, may be obtained by using Equation 4.16b in 4.17 simplifying

$$\begin{bmatrix} \Delta x_0 \\ \Delta y_0 \\ \Delta z_0 \end{bmatrix} \approx (diag(x_0 \;\; y_0 \;\; z_0))^{-1}(\mathbf{G}^T\mathbf{B}^{-T}\mathbf{B}^{-1}\mathbf{G})^{-1}\mathbf{G}^T\mathbf{B}^{-T} \begin{bmatrix} \Delta x_0 \\ \Delta y_0 \\ \Delta z_0 \\ \Delta r_0 \end{bmatrix} \tag{4.18}$$

We refer to $(diag(x_0 \;\; y_0 \;\; z_0))^{-1}(\mathbf{G}^T\mathbf{B}^{-T}\mathbf{B}^{-1}\mathbf{G})^{-1}\mathbf{G}^T\mathbf{B}^{-T}$ as an error-reducing matrix (3×4).

EXAMPLE 4.1

We show some properties of an error-reducing matrix. The matrix is independent of the scale factor, that is, it does not change even if we relocate the transmitter to a position whose coordinates are a constant multiple of its earlier location.

Error-reducing matrix: source at (10, 10, 10)

0.83333	−0.16667	−0.16667	0.28868
−0.16667	0.83333	−0.16667	0.28868
−0.16667	−0.16667	0.83333	0.28868

Assume that errors in the estimates of the transmitter location are 1% of the true values, that is, in earlier example $\Delta x_0 = \Delta y_0 = \Delta z_0 = \Delta r_0 = 0.1$. By imposing the constraint, $x_0^2 + y_0^2 + z_0^2 = r_0^2$; the errors in location are reduced to [0.09666, 0.09666, 0.083298]. The error-reducing matrix remains unchanged along a radial line but it varies with angle. For example, with the source on the x-axis, (10, 0.001, 0.001), the reducing matrix turns out

5.00e − 01	−5.00e − 05	−5.00e − 05	5.00e − 01
−5.00e − 05	1.00e + 00	−5.00e − 09	5.00e − 05
−5.00e − 05	−5.00e − 09	1.00e + 00	5.00e − 05

4.2 ToA MEASUREMENTS

4.2.1 Simple Method

Let τ_n, $n = 1,\ldots N$ be measured ToA at N anchor nodes located at known location (x_n, y_n), $n = 1,\ldots N$. Let a source (transmitter) be at (x_s, y_s). We then have the following set of equations:

$$(c\tau_n)^2 = \left\{(x_n - x_s)^2 + (y_n - y_s)^2\right\}, \quad n = 1,\cdots N \tag{4.19}$$

Equation 4.19 may be expressed in a different form, linear with respect to unknowns (x_s, y_s).

$$x_n x_s + y_n y_s = \frac{1}{2}(x_n^2 + y_n^2 + x_s^2 + y_s^2 - (c\tau_n)^2) \quad n = 1,\ldots N \tag{4.20}$$

To remove the quadratic terms in (x_s,y_s) we subtract, say, the first equation ($n=1$) from all other equations

$$\begin{aligned}&(x_n - x_1)x_s + (y_n - y_1)y_s\\ &= \frac{1}{2}[x_n^2 + y_n^2 - (c\tau)^2 - x_1^2 - y_1^2 + (c\tau)^2],\\ &n = 2,\ldots,N\end{aligned} \tag{4.21}$$

Note that the right-hand side of Equation 4.21 is a known quantity. It is a linear equation in (x_s,y_s). It can be readily solved to obtain a least-squares solution for (x_s,y_s)

$$\begin{bmatrix} x_2 - x_1 & y_2 - y_1 \\ x_3 - x_1 & y_3 - y_1 \\ \cdots & \\ x_N - x_1 & y_N - y_1 \end{bmatrix}\begin{bmatrix} x_s \\ y_s \end{bmatrix} = \frac{1}{2}\begin{bmatrix} x_2^2 + y_2^2 - (c\tau_2)^2 - x_1^2 - y_1^2 + (c\tau_1)^2 \\ x_3^2 + y_3^2 - (c\tau_3)^2 - x_1^2 - y_1^2 + (c\tau_1)^2 \\ \cdots \\ x_N^2 + y_N^2 - (c\tau_N)^2 - x_1^2 - y_1^2 + (c\tau_1)^2 \end{bmatrix} \tag{4.22a}$$

$$\mathbf{A}\begin{bmatrix} x_s \\ y_s \end{bmatrix} = \mathbf{B} \tag{4.22b}$$

The least-squares solution of Equation 4.22b is given by

$$\begin{bmatrix} x_s \\ y_s \end{bmatrix} = [\mathbf{A}^T\mathbf{A}]^{-1}\mathbf{A}^T\mathbf{B} \tag{4.22c}$$

It should be noted that the accuracy of the previous method, even in the absence of measurement noise, depends upon the rank condition of $[\mathbf{A}^{\mathrm{T}}\ \mathbf{A}]$. It is singular, for example, when all sensors are co-linear. It is also possible to select another sensor as a reference sensor and generate N−1 equations of the type shown in Equation 4.22a. In the absence of noise, however, all of them yield the same answer, but in the presence of measurement noise, we get different answers, giving additional diversity. Caffery [4] suggested this possibility.

4.2.2 Iterative Method

When the ToA estimates are noisy, noise-squared terms in Equation 4.22a may introduce bias in the estimated source coordinates. Chan [5] modified the above least square approach to remove the square terms. Consider the following cost function:

$$\varepsilon^2 = \sum_{n=1}^{N}\left(\sqrt{(x_n - x_s)^2 + (y_n - y_s)^2} - c\,\tau_n\right)^2 \tag{4.23}$$

We like to minimize the cost function with respect to (x_s,y_s). For this, the derivatives of ε^2 with respect to x_s and y_s are set to zero. We obtain

$$\frac{d\varepsilon^2}{dx_s} = \sum_{n=1}^{N} \frac{\sqrt{(x_n - x_s)^2 + (y_n - y_s)^2} - c\tau_n}{\sqrt{(x_n - x_s)^2 + (y_n - y_s)^2}}(x_n - x_s) = 0 \tag{4.24}$$

In Equation 4.24, ToAs (hence measurement noise) appear in linear form. Assuming low noise power (high SNR), we can approximate $c\tau_n \cong \sqrt{(x_n - x_s)^2 + (y_n - y_s)^2}$. Using this approximation, we can further write (using an identity),

$$(\mathrm{a} - \mathrm{b}) = (\mathrm{a}^2 - \mathrm{b}^2)/(\mathrm{a} + \mathrm{b})$$

as

$$\begin{aligned} &\sqrt{(x_n - x_s)^2 + (y_n - y_s)^2} - c\tau_n \\ &\cong \frac{(x_n - x_s)^2 + (y_n - y_s)^2 - (c\tau_i)^2}{2c\tau_i} \end{aligned} \tag{4.25}$$

Using Equation 4.25, in 4.24 we obtain, after some simplifications,

$$\begin{aligned} &\sum_{n=1}^{N} \frac{(r_s^2 - 2x_s x_n - 2y_s y_n)e_n + g_n}{2c\tau_n} = 0 \\ &\sum_{n=1}^{N} \frac{(r_s^2 - 2x_s x_n - 2y_s y_n)f_n + h_n}{2c\tau_n} = 0 \end{aligned} \tag{4.26}$$

where

$$r_s^2 = x_s^2 + y_s^2 \quad \text{(constraint)} \tag{4.27}$$

and

$$e_n = \frac{(x_n - x_s)}{\sqrt{(x_n - x_s)^2 + (y_n - y_s)^2}}, \quad f_n = \frac{(y_n - y_s)}{\sqrt{(x_n - x_s)^2 + (y_n - y_s)^2}}$$

$$g_n = (x_n^2 + y_n^2 - (c\tau_n)^2)e_n, \quad h_n = (x_n^2 + y_n^2 - (c\tau_n)^2)f_n$$

Equation 4.26 may be expressed in a matrix form

$$\begin{bmatrix} \sum_{n=1}^{N} \frac{x_n e_n}{c\tau_n} & \sum_{n=1}^{N} \frac{y_n e_n}{c\tau_n} \\ \sum_{n=1}^{N} \frac{x_n f_n}{c\tau_n} & \sum_{n=1}^{N} \frac{y_n f_n}{c\tau_n} \end{bmatrix} \begin{bmatrix} x_s \\ y_s \end{bmatrix} = \frac{1}{2} \begin{bmatrix} r_s^2 \sum_{n=1}^{N} \frac{e_n}{c\tau_n} + \sum_{n=1}^{N} \frac{g_n}{c\tau_n} \\ r_s^2 \sum_{n=1}^{N} \frac{f_n}{c\tau_n} + \sum_{n=1}^{N} \frac{h_n}{c\tau_n} \end{bmatrix} \tag{4.28}$$

Equation 4.28 cannot be solved in a closed form because of the presence of (x_s, y_s) in r_s^2, e_n, and f_n. Starting with the initial values $e_n = x_n/c\tau_n$ and $f_n = y_n/c\tau_n$, we need to solve Equation 4.28, subject to the constraint given by Equation 4.27. Let us express Equation 4.28 in a matrix form

$$\mathbf{S}\mathbf{x}_S = \boldsymbol{\delta}$$

$$\mathbf{x}_S = \mathbf{S}^{-1}\boldsymbol{\delta} \tag{4.29}$$

$$r_S^2 = \mathbf{x}_S^T\mathbf{x}_S = \boldsymbol{\delta}^T\mathbf{S}^{-T}\mathbf{S}^{-1}\boldsymbol{\delta} \tag{4.30a}$$

An iterative approach is possible, where starting from some initial values for (x_s, y_s) we generate new values for (x_s, y_s), which are then plugged into Equation 4.27, leading to a quadratic equation in r_s^2. To show this, let us express $\boldsymbol{\delta} = r_s^2\mathbf{a} + \mathbf{b}$ where

$$\mathbf{a} = \frac{1}{2}\left[\sum_{n=1}^{N}\frac{e_n}{c\tau_n} \quad \sum_{n=1}^{N}\frac{f_n}{c\tau_n}\right]$$

and

$$\mathbf{b} = \frac{1}{2}\left[\sum_{n=1}^{N}\frac{g_n}{c\tau_n} \quad \sum_{n=1}^{N}\frac{h_n}{c\tau_n}\right]^T$$

Equation 4.30a may be simplified as

$$r_s^2 = r_s^4\mathbf{a}^T\Sigma\mathbf{a} + r_s^2(\mathbf{b}^T\Sigma\mathbf{a} + \mathbf{a}^T\Sigma\mathbf{b}) + \mathbf{b}^T\Sigma\mathbf{b} \tag{4.30b}$$

where $\Sigma = \mathbf{S}^{-T}\,\mathbf{S}^{-1}$. The solution Equation 4.30b is easily obtained as

$$r_s^2 = \frac{-(\mathbf{b}^T\Sigma\mathbf{a} + \mathbf{a}^T\Sigma\mathbf{b} - 1) \pm \sqrt{(\mathbf{b}^T\Sigma\mathbf{a} + \mathbf{a}^T\Sigma\mathbf{b} - 1)^2 - 4\mathbf{a}^T\Sigma\mathbf{a}\,\mathbf{b}^T\Sigma\mathbf{b}}}{2\mathbf{a}^T\Sigma\mathbf{a}} \tag{4.31}$$

Since r_s^2 is a measure of distance, it must be a real positive quantity and hence we need to select a real positive root, out of two possible roots. When both roots are positive, we need to select a larger root when in far field and a smaller root when in near field. This estimate of the distance is next plugged into Equation 4.29, and we get a fresh estimate of the source location. This completes one iteration. The process is carried out over a few iterations, often four to five iterations are enough, to reach convergence (see Figure 4.3).

EXAMPLE 4.2

We consider a DSA consisting of four sensors and one transmitter located at $\mathbf{x}_s = [12\ \ 8]$ in meters. The shape of the array is sketched in Figure 4.2. The computed ToAs are [0.0095, 0.0055, 0.0019, 0.0081 (sec)] (assuming sound speed = 1500 m/sec in sea water). The ToA measurement errors were modeled as uniformly distributed random variables in the range $\pm\,0.00025$. To start with, we assume that the source was located at (0, 0). Ten iterations were performed.

The nature of the convergence of estimated source location is shown in Figure 4.3. About four to five iterations were enough to reach convergence

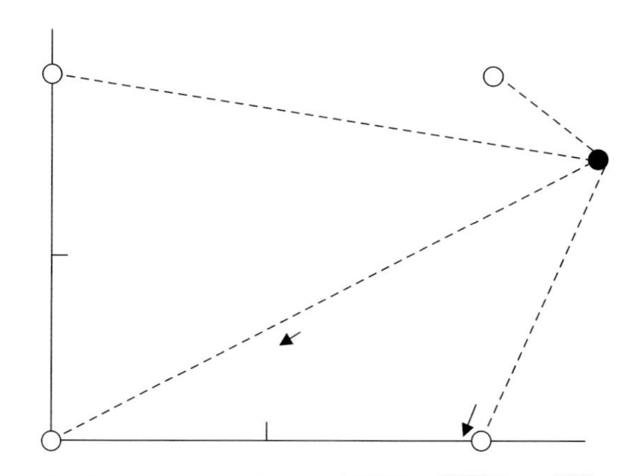

FIGURE 4.2 Shape of the array and position of the source.

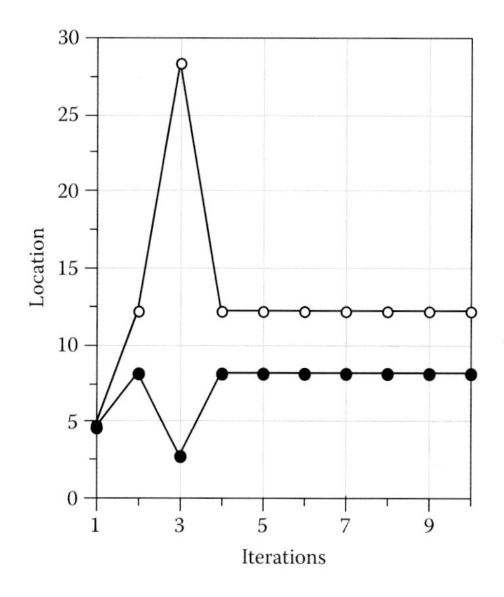

FIGURE 4.3 Iterative solution of non-linear equation for ToA estimation. Convergence is plotted as shown: (○) x-coordinate and (•) y-coordinate.

EXAMPLE 4.3

We have yet another example of transmitter localization with a DSA. We have eight sensors whose x- and y-coordinates in meters are [100, 400; 200, 200; 200, 300; 400, 100; 400, 200; 500, 400; 300, 700; 250, 800]. The geometry of the array is shown in Figure 4.4. The transmitter is placed at (1000, 40) meters. The wave speed is 3.0*10^8 m/sec or 30 cm/nsec (speed of radio waves). We model the errors in the ToA estimates as uniformly distributed random variables lying in the range ±Q nanoseconds We have used both least-squares and iterative algorithms for localization using noisy ToA estimates. Only five iterations were carried out in the latter algorithm. Mean and variance of the estimated x- and y-coordinates of the transmitter were computed from the results obtained from one hundred independent trials. The results are shown in Tables 4.1 and 4.2. As expected the mean square error (MSE) in the location estimate increases with increasing error in ToA estimates, but this error is less in the iterative method.

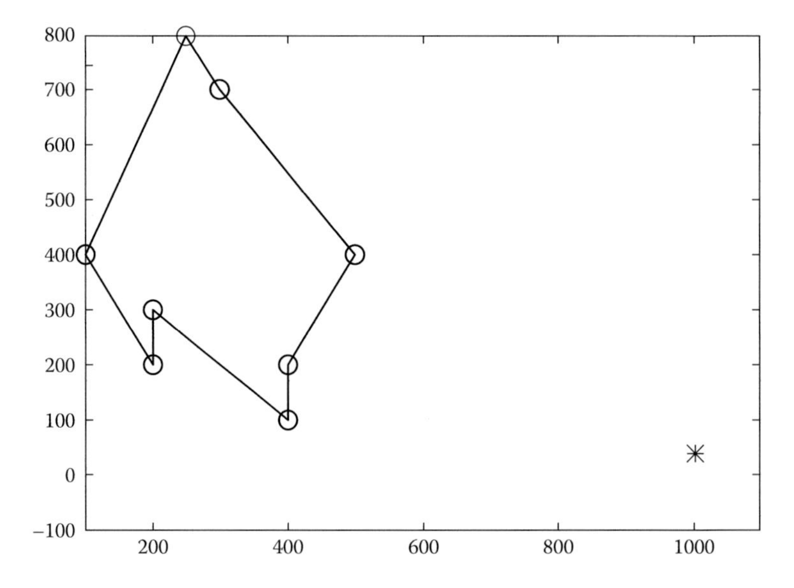

FIGURE 4.4 Eight-sensor DSA. The position of the transmitter is shown with "*" and the estimated position with "*". They appear to be overlapping, as the error is very small.

TABLE 4.1
Mean and Variance of Estimated x- and y-Coordinates with a Least-Squares Algorithm

	Q = 10	15	25
Mean of x-coordinate	−0.062	0.4584	0.6082
Mean of y-coordinate	0.0282	0.2891	−0.1594
Variance of x	15.54	30.939	112.12
Variance of y	5.22	5.375	39.62

Note: ToA errors are modeled as a uniformly distributed random variable in the range ± Q nanoseconds.

TABLE 4.2
Mean and Variance of Estimated x- and y-Coordinates with Iterative (Five Iterations) Algorithm

	Q = 10	15	25
Mean of x-coordinate	−0.0357	−0.1239	−0.1269
Mean of y-coordinate	−0.0084	−0.3704	0.0890
Variance of x	1.172	3.797	11.463
Variance of y	4.669	12.78	37.559

By increasing the number of iterations does not seem to improve the estimates significantly. However, the estimates appear to be unbiased particularly for small error in ToA estimates.

4.3 TDoA MEASUREMENTS

4.3.1 Re-Minimization

Consider a single transmitter and a distributed sensor (M sensors) array. The TDoA between the mth and nth sensors is defined as

$$\tau_{mn} = \tau_m - \tau_n$$

which is actually the measured quantity. Let $\tau_{mn}(\hat{\mathbf{x}}_s)$ be TDoA for $\hat{\mathbf{x}}_s$, the assumed location of the transmitter. We need to minimize the mean square difference (output error) between $(\tau_{mn} - \tau_{mn}(\hat{\mathbf{x}}_s))$ with respect to $\hat{\mathbf{x}}_s$

$$(\tau_{mn} - \tau_{mn}(\hat{\mathbf{x}}_s))^2 = \min \quad \text{w.r.t} \quad \hat{\mathbf{x}}_s \tag{4.32a}$$

Minimization of Equation 4.32a is highly non-linear and also non-convex, that is, there are multiple minima. We shall express Equation 4.32a in a different form. Note that $d_{mn} = c\tau_{mn} = r_{sn} - r_{sm}$ and

$$r_{sn} = \sqrt{r_{sm}^2 + r_{mn}^2 - 2((x_s - x_n)(x_m - x_s) + (y_s - y_n)(y_m - y_s))}$$

Following geometry shown in Figure 4.5 we can write

$$(r_{sm} + d_{mn})^2 = r_{sm}^2 + r_{mn}^2 - 2((x_s - x_n)(x_m - x_n) + (y_s - y_n)(y_m - y_n))$$

After simplifying the previous equation, we get

$$0 = \begin{bmatrix} r_{mn}^2 - 2r_{sm}d_{mn} - d_{mn}^2 \\ -2((x_s - x_n)(x_m - x_n) + (y_s - y_n)(y_m - y_n)) \end{bmatrix} \tag{4.32b}$$

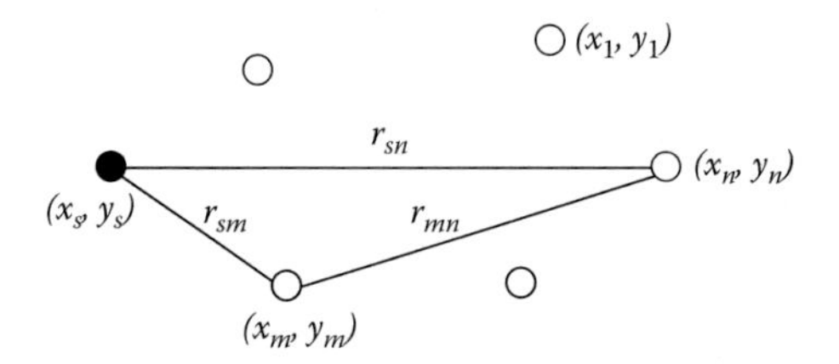

FIGURE 4.5 DSA of five sensors (O) and one transmitter (●). r_{sm} stands for the distance between the transmitter and the *m*th sensor, and r_{mn} stands for the distance between the *m*th and the *n*th sensors.

where r_{mn} is distance between *m*th and *n*th sensor, that is, $r_{mn} = \sqrt{(x_m - x_n)^2 + (y_m - y_n)^2}$. In the presence of measurement errors, in τ_{mn} the left-hand side of Equation 4.32b will not be zero but a finite random quantity, ε_{mn}, termed as "equation error" [6, 19]. Surprisingly, the "equation error" is linear in unknown parameters (x_s, y_s).

$$\varepsilon_{mn} = \begin{bmatrix} r_{mn}^2 - 2r_{sm}d_{mn} - d_{mn}^2 \\ \\ -2((x_s - x_n)(x_m - x_n) + (y_s - y_n)(y_m - y_n)) \end{bmatrix} \tag{4.33}$$

Consider a simple situation where we have TDoAs with reference to just one sensor, say m = 1. Define the following vectors and matrices:

$$\boldsymbol{\varepsilon} = [\varepsilon_{12}, \varepsilon_{13}, \cdots, \varepsilon_{1M}]$$

$$\boldsymbol{\delta} = [r_{12}^2 - d_{12}^2, r_{13}^2 - d_{13}^2, \cdots, r_{1M}^2 - d_{1M}^2]^T$$

$$\mathbf{d} = [d_{12}, d_{13}, \cdots, d_{1M}]^T$$

$$\mathbf{T} = [x_1(x_1 - x_2) + y_1(y_1 - y_2), x_1(x_1 - x_3) + y_1(y_1 - y_3), \cdots, x_1(x_1 - x_M) + y_1(y_1 - y_M)]$$

and

$$\mathbf{S} = \begin{bmatrix} x_2 - x_1 & y_2 - y_1 \\ x_3 - x_1 & y_3 - y_1 \\ \vdots & \vdots \\ x_M - x_1 & y_M - y_1 \end{bmatrix}$$

Express Equation 4.33, using the previous quantities, as follows

$$\boldsymbol{\varepsilon} = \boldsymbol{\delta} - 2r_{s1}\mathbf{d} - 2\mathbf{S}\mathbf{x}_s - 2\mathbf{T} \tag{4.34}$$

where

$$\mathbf{x}_s = [x_s, y_s]^T$$

and

$$r_{s1} = \sqrt{(x_s - x_1)^2 + (y_s - y_1)^2}$$

By varying m from 2 to M (and counting each $d_{mn} = d_{nm}$ pair once) we get $(M^2-M)/2$ equations of the type Equation 4.34. However, in the noiseless case, any $(M-1)$ measurements, which form a "minimal spanning sub-tree," can determine all the rest [6]. However, when there is noise, the previously mentioned redundancy can be exploited to increase noise immunity [7]. Formally, by minimizing the equation error (power), we obtain the result

$$\mathbf{S}\hat{\mathbf{x}}_s = \frac{1}{2}(\boldsymbol{\delta} - 2r_{s1}\mathbf{d} - 2\mathbf{T}) \tag{4.35a}$$

The least-squares solution for $\hat{\mathbf{x}}_s$ is given by

$$\hat{\mathbf{x}}_s = \frac{1}{2}[\mathbf{S}\mathbf{S}^T]^{-1}\mathbf{S}^T(\boldsymbol{\delta} - 2r_{s1}\mathbf{d} - 2\mathbf{T}) \tag{4.35b}$$

$$= \frac{1}{2}\mathbf{S}^{\dagger}(\boldsymbol{\delta} - 2r_{s1}\mathbf{d} - 2\mathbf{T})$$

where $\mathbf{S}^{\dagger}$ is the pseudo-inverse of $\mathbf{S}$. We still need to obtain an estimate of r_{s1}. Smith [6] suggested an approach involving re-minimization of Equation 4.34 with respect to r_{s1} after substituting the unknown transmitter coordinates with the estimated ones obtained from Equation 4.35. The equation error reduces to

$$\boldsymbol{\varepsilon} = \boldsymbol{\delta} - \mathbf{2}r_{s1}\mathbf{d} - \mathbf{S}\mathbf{S}^{\dagger}(\boldsymbol{\delta} - \mathbf{2}r_{s1}\mathbf{d} - 2\mathbf{T}) - 2'$$

$$= (\mathbf{I} - \mathbf{S}\mathbf{S}^{\dagger})(\boldsymbol{\delta} - \mathbf{2}r_{s1}\mathbf{d} - 2\mathbf{T}) \tag{4.36}$$

and squared error is given by

$$\|\boldsymbol{\varepsilon}\|^2 = (\boldsymbol{\delta} - \mathbf{2}r_{s1}\mathbf{d} - 2\mathbf{T})^T(\mathbf{I} - \mathbf{S}\mathbf{S}^{\dagger})(\boldsymbol{\delta} - \mathbf{2}r_{s1}\mathbf{d} - 2\mathbf{T} \tag{4.37}$$

We shall now minimize the squared error with respect to r_{s1}. The result is

$$\hat{r}_{s1} = \frac{\mathbf{d}^T(\mathbf{I} - \mathbf{S}\mathbf{S}^{\dagger})(\boldsymbol{\delta} - 2\mathbf{T})}{2\mathbf{d}^T(\mathbf{I} - \mathbf{S}\mathbf{S}^{\dagger})\mathbf{d}} \tag{4.38}$$

Substituting for r_{s1} in Equation 4.35 we obtain

$$\hat{\mathbf{x}}_s = \frac{1}{2}\mathbf{S}^{\dagger}[\mathbf{I} - \frac{\mathbf{d}\mathbf{d}^T(\mathbf{I}-\mathbf{S}\mathbf{S}^{\dagger})}{2\mathbf{d}^T(\mathbf{I}-\mathbf{S}\mathbf{S}^{\dagger})\mathbf{d}}](\boldsymbol{\delta} - 2\mathbf{T}) \tag{4.39}$$

We shall now relate the "equation error" to the output error in TDoA. For this, add and subtract $\hat{r}_{s1}^2$ in Equation 4.33 for $m = 1$. After simplification, we obtain

$$\varepsilon_n = r_{1n}^2 - 2((x_1 - x_s)(x_1 - x_n) + (y_1 - y_s)(y_1 - y_n)) + \hat{r}_{s1}^2$$

$$-(d_{1n}^2 + 2r_{s1}d_{1n} + \hat{r}_{s1}^2)$$

$$\varepsilon_n = \left\|\mathbf{x}_n - \hat{\mathbf{x}}_s\right\|^2 - (\hat{r}_{s1} + d_{1n})^2 \tag{4.40}$$

where $\mathbf{x}_n$ is the location vector of the nth sensor (hats denote estimated quantities). Further, the distance to the nth sensor with respect to the transmitter may be expressed as (see Figure 4.6)

$$\left\|\mathbf{x}_n - \hat{\mathbf{x}}_s\right\| = (\hat{r}_{s1} + \hat{d}_{1n}) \tag{4.41}$$

Using Equation 4.41 in 4.40 we get

$$\varepsilon_n = (\hat{r}_{s1} + \hat{d}_{1n})^2 - (\hat{r}_{s1} + d_{1n})^2$$

$$= (\hat{d}_{1n} + 2\hat{r}_{s1} + d_{1n})(\hat{d}_{1n} - d_{1n}) \tag{4.42}$$

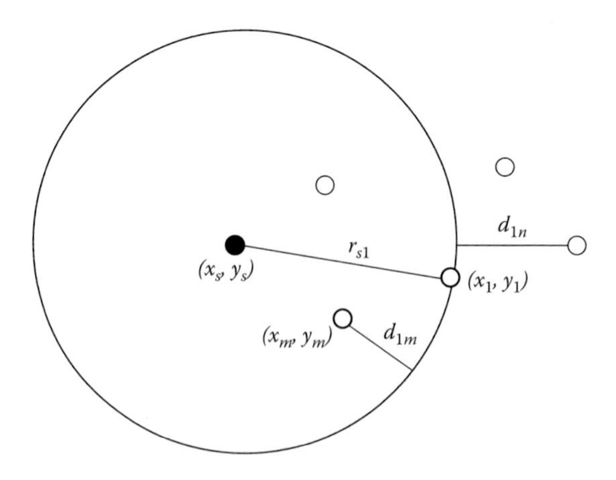

FIGURE 4.6 A circle is drawn centered at an unknown transmitter position and passing through the first sensor, so as to minimize the sum of squares of the radial distance to the remaining sensors with respect to the circle.

Assuming the noise in the TDoA measurement is small, that is, $\hat{d}_{1n} \approx d_{1n}$, Equation 4.42 may be approximated as

$$\begin{aligned}\varepsilon_n &\approx 2(\hat{r}_{s1} + d_{1n})(\hat{d}_{1n} - d_{1n}) \\ &= 2\hat{r}_{sn}(\hat{d}_{1n} - d_{1n})\end{aligned} \tag{4.43}$$

Note that $(\hat{d}_{1n} - d_{1n}) = c(\hat{\tau}_{1n} - \tau_{1n})$, that is, the propagation speed multiplied by error in the TDoA measurement. The distance to the transmitter also influences the "equation error." Indeed the "equation error" increases linearly with the distance to transmitter.

4.3.2 Treat Range as a Variable

In Equation 4.34, we have treated x_s, y_s, and r_{s1} as three independent unknowns even though they are all related through a relation,

$$r_{s1} = \sqrt{(x_s - x_1)^2 + (y_s - y_1)^2}$$

Equation 4.34 reduces to

$$\boldsymbol{\varepsilon} = \delta - \mathbf{S}'(\mathbf{x}_s^T, r_{s1})^T - 2\mathbf{T} \tag{4.44}$$

where

$$\mathbf{S}' = \begin{bmatrix} x_2 - x_1 & y_2 - y_1 & d_{12} \\ x_3 - x_1 & y_3 - y_1 & d_{13} \\ \vdots & \vdots & \\ x_M - x_1 & y_M - y_1 & d_{1M} \end{bmatrix}$$

Solving Equation 4.44 under the assumption that the equation error is a white noise process with uniform variance, we obtain

$$\begin{bmatrix} \hat{x}_s \\ \hat{y}_s \\ \hat{r}_{s1} \end{bmatrix} = \frac{1}{2}\mathbf{S}'^{\dagger}[\boldsymbol{\delta} - 2\mathbf{T}] \tag{4.45}$$

where $\mathbf{S}'^{\dagger}$ is the pseudo-inverse of $\mathbf{S}'$. Note that the errors in $\hat{x}_s, \hat{y}_s, \hat{r}_{s1}$ are without any constraint on the range.

We shall next impose the range constraint and reestimate the transmitter location. For this, subtract the reference sensor coordinates from the estimates and square each element of the resulting vector,

$$(\hat{x}_s - x_1)^2 = (x_s^0 + \varepsilon_1 - x_1)^2$$

$$(\hat{y}_s - y_1)^2 = (y_s^0 + \varepsilon_2 - y_1)^2$$

$$\hat{r}_{s1}^2 = (r_{s1}^0 + \varepsilon_3)^2$$

where x_s^0, y_s^0 are actual transmitter location coordinates and r_{s1}^0 is the actual range to the transmitter from the reference sensor, and ε_1 ε_2 and ε_3 are estimation errors without a range constraint. Now we impose the range constraint. The resulting error vector, after ignoring quadratic error terms (i.e., $\varepsilon_1^2, \varepsilon_2^2, \varepsilon_3^2$), is given as

$$\tilde{\boldsymbol{\varepsilon}} = 2\begin{bmatrix} (x_s^0 - x_1)\varepsilon_1 \\ (y_s^0 - y_1)\varepsilon_2 \\ r_{s1}^0\varepsilon_3 \end{bmatrix} = \mathbf{h} - \mathbf{G}\begin{bmatrix} (x_s^0 - x_1)^2 \\ (y_s^0 - y_1)^2 \end{bmatrix} \tag{4.46}$$

where

$$\mathbf{h} = \begin{bmatrix} (\hat{x}_s - x_1)^2 \\ (\hat{y}_s - y_1)^2 \\ (r_{s1}^0 + \varepsilon_3)^2 \end{bmatrix} \text{and } \mathbf{G} = \begin{bmatrix} 1 & 0 \\ 0 & 1 \\ 1 & 1 \end{bmatrix}$$

Notice that the new error vector $\tilde{\boldsymbol{\varepsilon}}$ is due to a non-zero correlation in $\left[\hat{x}_s, \hat{y}_s, \hat{r}_{s1}\right]^T$ on the left-hand side of Equation 4.45). It is a scaled version of ε Equation 4.44. In particular, the covariance matrix of errors (zero mean) is given by

$$\mathbf{C}_{\tilde{\varepsilon}} = E\left\{\tilde{\boldsymbol{\varepsilon}}\tilde{\boldsymbol{\varepsilon}}^T\right\} = 4\mathbf{B}\Sigma\mathbf{B}^T \tag{4.47}$$

where

$$\mathbf{B} = diag\left\{(x_s^0 - x_1), (y_s^0 - y_1), r_{s1}^0\right\}$$

$$\boldsymbol{\Sigma} = \mathbf{E}\{\boldsymbol{\varepsilon}\boldsymbol{\varepsilon}^T\}$$

A weighted least-squares (WLS) solution of Equation 4.46, using the inverse covariance matrix given in Equation 4.47 as a weighting matrix, is given in [8]. For this, we minimize weighted mean square error with respect to **h**.

$$\mathbf{E}\left\{\tilde{\boldsymbol{\varepsilon}}^T\mathbf{C}_{\tilde{\boldsymbol{\varepsilon}}}^{-1}\tilde{\boldsymbol{\varepsilon}}\right\} = E\left\{\begin{array}{l} \mathbf{h}^T\mathbf{C}_{\tilde{\boldsymbol{\varepsilon}}}^{-1}\mathbf{h} + \begin{bmatrix}(x_s^0 - x_1)^2 \\ (y_s^0 - y_1)^2\end{bmatrix}^T + \mathbf{G}^T\mathbf{C}_{\tilde{\boldsymbol{\varepsilon}}}^{-1}\mathbf{G}\begin{bmatrix}(x_s^0 - x_1)^2 \\ (y_s^0 - y_1)^2\end{bmatrix} \\ -\mathbf{h}^T\mathbf{C}_{\tilde{\boldsymbol{\varepsilon}}}^{-1}\mathbf{G}\begin{bmatrix}(x_s^0 - x_1)^2 \\ (y_s^0 - y_1)^2\end{bmatrix} - \begin{bmatrix}(x_s^0 - x_1)^2 \\ (y_s^0 - y_1)^2\end{bmatrix}^T \mathbf{G}^T\mathbf{C}_{\tilde{\boldsymbol{\varepsilon}}}^{-1}\mathbf{h} \end{array}\right.$$

where E{ } stands for expected operation. Differentiate the weighted mean square error with respect to **h** and set the derivative to zero. Note that the second term is simply a constant. We obtain an equation for location coordinates that minimizes the mean square error:

$$\begin{bmatrix}(\hat{x}_s' - x_1)^2 \\ (\hat{y}_s' - y_1)^2\end{bmatrix} = (\mathbf{G}^T\mathbf{C}_{\tilde{\boldsymbol{\varepsilon}}}^{-1}\mathbf{G})^{-1}(\mathbf{G}^T\mathbf{C}_{\tilde{\boldsymbol{\varepsilon}}}^{-1}\mathbf{h}) \tag{4.48}$$

where $\hat{x}_s'$ and $\hat{y}_s'$ are the WLS estimates with range constraint. The inverse of $\mathbf{C}_{\tilde{\boldsymbol{\varepsilon}}}$, when $\boldsymbol{\varepsilon}$ is assumed as an uncorrelated random vector, is readily obtained as

$$\begin{aligned}\mathbf{C}_{\tilde{\boldsymbol{\varepsilon}}}^{-1} &= \frac{1}{4}\mathbf{B}^{-1}\Sigma^{-1}\mathbf{B}^{-T} \\ &= \frac{1}{4}diag\left\{\frac{1}{\sigma_{\varepsilon_1}^2(x_s^0 - x_1)^2}\ \frac{1}{\sigma_{\varepsilon_2}^2(y_s^0 - y_1)^2}\ \frac{1}{\sigma_{\varepsilon_3}^2(r_{s1}^0)^2}\right\}\end{aligned}$$

where $\sigma_{\varepsilon_1}^2$, $\sigma_{\varepsilon_2}^2$ and $\sigma_{\varepsilon_3}^2$ are variances of ε_1, ε_2, and ε_3, respectively.

4.3.3 Annihilation of Range Variable

There is yet another approach where in Equation 4.34 we annihilate a term, which depends on the transmitter range, namely, $r_{s1}\mathbf{d}$. The idea is simply to devise an annihilating matrix **P,** such that $\mathbf{Pd} = 0$ [9]. Define a matrix as

$$\mathbf{D} = \begin{bmatrix} d_{12} & & & \\ & d_{13} & & \\ & & \ddots & \\ & & & d_{1M} \end{bmatrix}^{-1}_{(M-1)\times(M-1)}$$

and a circular shift matrix as

$$\mathbf{Z} = \begin{bmatrix} 0 & 1 & 0 & \cdots & 0 \\ 0 & 0 & 1 & \cdots & 0 \\ \vdots & & & \ddots & \\ 0 & 0 & 0 & \ddots & 1 \\ 1 & 0 & 0 & \ldots & 0 \end{bmatrix}_{(M-1)\times(M-1)}$$

Define an annihilating matrix as $\mathbf{P} = (\mathbf{I} - \mathbf{Z})\mathbf{D}$. By actual multiplication, we can show that $\mathbf{Pd} = \mathbf{0}$. Pre-multiply Equation 4.34 with $\mathbf{P}$:

$$\mathbf{P}\boldsymbol{\varepsilon} = \mathbf{P}\boldsymbol{\delta} - 2\mathbf{PSx}_s - 2\mathbf{PT} \tag{4.49}$$

The error term on the left-hand side now becomes

$$\mathbf{P}\boldsymbol{\varepsilon} = \left[\frac{\varepsilon_{12}}{d_{12}} - \frac{\varepsilon_{13}}{d_{13}}, \frac{\varepsilon_{13}}{d_{13}} - \frac{\varepsilon_{14}}{d_{14}}, \ldots, \frac{\varepsilon_{1M}}{d_{1M}} - \frac{\varepsilon_{12}}{d_{12}}\right]^T$$

The least mean square solution of Equation 4.49 is given by

$$\hat{\mathbf{x}}_s = \frac{1}{2}(\mathbf{S}^T\mathbf{P}^T\mathbf{PS})^{-1}(\mathbf{PS})^T(\mathbf{P}\boldsymbol{\delta} - 2\mathbf{PT}) \tag{4.50}$$

We assume that $(\mathbf{S}^T\mathbf{P}^T\mathbf{PS})$ is invertible (full rank) although $\mathbf{P}$ is singular (rank = M−2). This would require the number of rows of $\mathbf{S}$ be greater than the number of columns, for example, for the planar array there are two columns, hence, M > 3.

4.3.4 Direction Cosines

Consider an alternate approach [10] where we estimate the direction cosines of the transmitter measured from a reference sensor. This does not require the range information. Let $\mathbf{r}_1, r_2, \ldots, \mathbf{r}_M$ represent distance vectors to M sensors and $\mathbf{r}_s$ the distance vector to the transmitter measured from the center of coordinates. TDoA at sensor m with reference to sensor 1 (reference sensor) may be expressed as

$$\tau_m - \tau_1 = \frac{|(\mathbf{r}_m - \mathbf{r}_s)|}{c} - \frac{|(\mathbf{r}_1 - \mathbf{r}_s)|}{c}$$

Note that τ_m represents travel time from the transmitter to the mth sensor. Rewriting the previous equation in a different form, we get

$$\tau_m - \tau_1 + \frac{|(\mathbf{r}_s - \mathbf{r}_1)|}{c} = \frac{|(\mathbf{r}_m - \mathbf{r}_1) - (\mathbf{r}_s - \mathbf{r}_1)|}{c} \tag{4.51a}$$

Squaring both sides of the previous equation, we obtain

$$(\tau_m - \tau_1)^2 + 2(\tau_m - \tau_1)\frac{|\mathbf{r}_s - \mathbf{r}_1|}{c}$$

$$= \left(\frac{|(\mathbf{r}_m - \mathbf{r}_1)|}{c}\right)^2 - 2\frac{(\mathbf{r}_m - \mathbf{r}_1).(\mathbf{r}_s - \mathbf{r}_1)}{c^2} \tag{4.51b}$$

Dividing both sides of Equation 4.51b by $|\mathbf{r}_s - \mathbf{r}_1|$ and rearranging the terms, we obtain

$$\tau_m - \tau_1 = -\frac{c(\tau_m - \tau)^2}{2|\mathbf{r}_s - \mathbf{r}_1|} + \frac{|(\mathbf{r}_s - \mathbf{r}_1)|^2}{2c|\mathbf{r}_s - \mathbf{r}_1|} - \frac{(\mathbf{r}_m - \mathbf{r}_1)(\mathbf{r}_s - \mathbf{r}_1)}{c|\mathbf{r}_s - \mathbf{r}_1|} \tag{4.52}$$

Define the following quantities, which are unknown:

$$\overline{\mathbf{x}}_s = (\overline{x}_s, \overline{y}_s) = \frac{(\mathbf{r}_s - \mathbf{r}_1)}{c|\mathbf{r}_s - \mathbf{r}_1|}$$

$$x_3 = \frac{1}{c|\mathbf{r}_s - \mathbf{r}_1|}, \quad \text{and} \quad x_4 = \frac{c}{2|\mathbf{r}_s - \mathbf{r}_1|}$$

Observe that $(\mathbf{r}_s - \mathbf{r}_1)/\,\mathbf{r}_s - \mathbf{r}_1$ represents the direction of cosines measured from the reference sensor to the transmitter. The direction cosines are independent of range, r_{s1}. Using quantities in Equation 4.52 we can express it as a system of linear equations in four unknowns: $\overline{x}_s, \overline{y}_s\, x_3$, and x_4:

$$(\tau_m - \tau) = -(\mathbf{r}_m - \mathbf{r}_1)\overline{\mathbf{x}}_s + |(\mathbf{r}_m - \mathbf{r}_1)|^2 x_3 - (\tau_m - \tau_1)^2 x_4 \tag{4.53}$$

We get M−1 linear equations, which may all be combined into a matrix form

$$\mathbf{Ax} = \mathbf{B} \tag{4.54a}$$

$$\mathbf{A} = \begin{bmatrix} -(\mathbf{r}_2 - \mathbf{r}_1) & |\mathbf{r}_2 - \mathbf{r}_1|^2 & -(\tau_2 - \tau_1)^2 \\ -(\mathbf{r}_3 - \mathbf{r}_1) & |(\mathbf{r}_3 - \mathbf{r}_1)|^2 & -(\tau_3 - \tau_1)^2 \\ \cdots & \cdots & \cdots \\ -(\mathbf{r}_M - \mathbf{r}_1) & |(\mathbf{r}_M - \mathbf{r}_1)|^2 & -(\tau_M - \tau_1)^2 \end{bmatrix}_{(M-1)\times 4}$$

$$\mathbf{x} = \begin{bmatrix} \overline{x}_s \\ \overline{y}_s \\ x_3 \\ x_4 \end{bmatrix} \quad \text{and} \quad \mathbf{B} = \begin{bmatrix} \tau_2 - \tau_1 \\ \tau_3 - \tau_1 \\ \vdots \\ \tau_M - \tau_1 \end{bmatrix}$$

Since there are four unknowns, we shall need at least five sensors (the reference sensor does not yield an equation), provided **A** is full rank, to solve for four unknowns. Four sensors will be enough if the propagation speed is known. However, in order to combat the presence of noise, we may need many more sensors. Then, we shall have more equations than unknowns, requiring the least square approach to solve for the unknowns. Unknown vector **x** is estimated by inverting Equation 4.54a:

$$\mathbf{x} = (\mathbf{A}^T\mathbf{A})^{-1}\mathbf{A}^T\mathbf{B} \tag{4.54b}$$

The emitter coordinates and the propagation speed can be derived from the estimated x,

$$\mathbf{x}_s = \frac{\overline{\mathbf{x}}_s}{2x_3} + \mathbf{r}_1 \quad \text{and} \quad c = \sqrt{\frac{x_4}{x_3}} \tag{4.55}$$

In some applications, for example, underwater or seismic applications, the propagation speed is not known a priori. Then, the previously mentioned method will enable us to estimate the unknown speed as well.

4.4 FDoA MEASUREMENTS

In this section, we explore how FDoA data can be used for localization. In order to generate FDoA, the transmitter/sensor must be moving. But, unlike ToA or TDoA measurements, there is no need for time synchronization in FDoA measurements. We begin with a simple illustration. We consider a 14-sensor distributed array (stationary on the ocean bottom) and a transmitter moving along the x-axis at a constant speed of 10 m/s. The transmitter is assumed to be at a height of 100 m above plane of array. The relative Doppler shift computed at different sensors is shown in Figure 4.7. A sensor, which lies overhead, records the lowest Doppler shift, that is, zero in the present case. All Doppler shifts at sensors in front of the emitter are negative and those behind are positive. This was also observed in Figure 2.10 when we projected all sensors and transmitters on to the x-axis. When the transmitter position is known, obtaining all three components of the emitter motion is straightforward. From Equation 2.9d, the Doppler shifts observed at M sensors are governed by M linear equations:

$$\frac{\delta f_m}{f_c} = \left(\frac{x_s(t) - x_m}{r_{sm}} u_x + \frac{y_s(t) - y_m}{r_{sm}} u_y + \frac{z_s(t)}{r_{sm}} u_z \right) \Big/ c \tag{4.56}$$

$$m = 1, \cdots, M$$

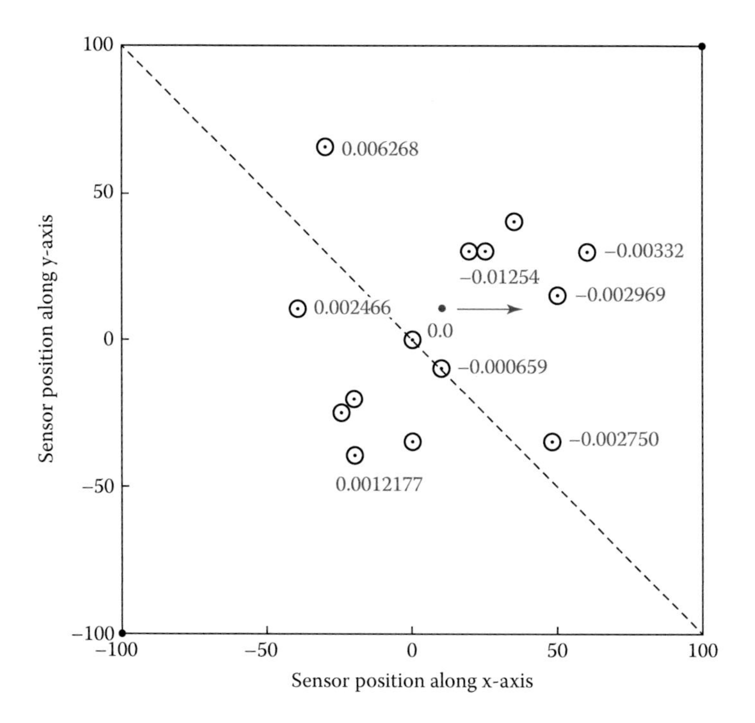

FIGURE 4.7 Relative Doppler shifts, $\delta f/f_c$, are observed at different sensors of a DSA of 14 sensors. The transmitter (red dot) is moving at a constant speed of 10 m/sec, along the x-axis. The array is presumed to be on the ocean bottom at a depth of 100 m below the transmitter. Some of the computed Doppler shifts are shown against each sensor.

where u_x, u_y, and u_z are three components of a moving transmitter, $x_s(t)$, $y_s(t)$, $z_s(t)$ are coordinates of the transmitter at time t and r_{sm} is the range of the transmitter s from the mth sensor. Define the following vectors and matrices:

$$\boldsymbol{\delta}\mathbf{f} = \left[\delta f_1, \quad \delta f_2, \quad \cdots \quad \delta f_M\right]^T$$

$$\mathbf{r} = diag\left\{{}^{1}\!/_{r_{s1}}, \quad {}^{1}\!/_{r_{s2}}, \quad \cdots \quad {}^{1}\!/_{r_{sM}}\right\}$$

$$\mathbf{S} = \begin{bmatrix} (x_s(t)-x_1) & (y_s(t)-y_1) & z_s(t) \\ \\ (x_s(t)-x_2) & (y_s(t)-y_2) & z_s(t) \\ \\ & \cdots & \\ \\ (x_s(t)-x_M) & (y_s(t)-y_M) & z_s(t) \end{bmatrix}$$

$$\mathbf{u}_s = \left[u_x, \quad u_y, \quad u_z\right]^T$$

Using the previous quantities, we can rewrite Equation 4.56 in matrix form

$$\delta \mathbf{f} = \frac{f_c}{c} \mathbf{rSu}_s \tag{4.57}$$

The least-squares solution of Equation 4.57 is straightforward, given by

$$\mathbf{u}_s = \frac{c}{f_c} (\mathbf{S}^T \mathbf{r}^T \mathbf{r} \ \mathbf{S})^{-1} (\mathbf{r} \ \mathbf{S})^T \delta \mathbf{f} \tag{4.58}$$

The rank status of $(\mathbf{S}^T \mathbf{r}^T \mathbf{rS})$ is important. We must have at least three ($M \geq 3$) non-collinear sensors in each subarray. Let us partition the sensor array into two non-overlapping subarrays, each having at least three sensors. For example, in Figure 4.7, a dashed line creates two such subarrays. The subarrays must have an equal number of sensors. Doppler shifts measured in each sub array will independently satisfy Equation 4.57. The two equations are

$$\delta \mathbf{f}_1 = \frac{f_c}{c} \mathbf{r}_1 \mathbf{S}_1 \mathbf{u}_s$$

$$\delta \mathbf{f}_2 = \frac{f_c}{c} \mathbf{r}_2 \mathbf{S}_2 \mathbf{u}_s$$

where subscripts 1 and 2 denote subarray one and subarray two, respectively. We shall use the first equation to solve for $\mathbf{u}_s$

$$\mathbf{u}_s = \frac{c}{f_c} (\mathbf{S}_1^T \mathbf{r}_1^T \mathbf{r}_1 \mathbf{S}_1)^{-1} (\mathbf{r}_1 \mathbf{S}_1)^T \delta \mathbf{f}_1 \tag{4.59a}$$

and use it in the second equation

$$\delta \mathbf{f}_2 = (\mathbf{r}_2 \mathbf{S}_2)(\mathbf{S}_1^T \mathbf{r}_1^T \mathbf{r}_1 \mathbf{S}_1)^{-1} (\mathbf{r}_1 \mathbf{S}_1)^T \delta \mathbf{f}_1 \tag{4.59b}$$

Let us express

$$\delta \mathbf{f}_2 = diag\left[\delta \mathbf{f}_2 . / \delta \mathbf{f}_1\right] \delta \mathbf{f}_1$$

in Equation 4.59b and obtain

$$diag\left\{\delta \mathbf{f}_2 ./ \delta \mathbf{f}_1\right\} \delta \mathbf{f}_1 = (\mathbf{r}_2 \mathbf{S}_2)(\mathbf{S}_1^T \mathbf{r}_1^T \mathbf{r}_1 \mathbf{S}_1)^{-1} (\mathbf{r}_1 \mathbf{S}_1)^T \delta \mathbf{f}_1 \tag{4.60}$$

Note that we have used a MATLAB convention that $\cdot/$ stands for an element-by-element division (assuming no element of $\delta \mathbf{f}_1$ is zero). We shall express Equation 4.60 in a standard eigen equation form:

$$d\mathbf{f}_1 = \left[diag\left\{d\mathbf{f}_1./d\mathbf{f}_2\right\}(\mathbf{r}_2\mathbf{S}_2)(\mathbf{S}_1^T\mathbf{r}_1^T\mathbf{r}_1\mathbf{S}_1)^{-1}(\mathbf{r}_1\mathbf{S}_1)^T\right]d\mathbf{f}_1$$

$$= \mathbf{\Gamma}d\mathbf{f}_1$$

where

$$\mathbf{\Gamma} = \left[diag\left\{\delta\mathbf{f}_1./\delta\mathbf{f}_2\right\}(\mathbf{r}_2\mathbf{S}_2)(\mathbf{S}_1^T\mathbf{r}_1^T\mathbf{r}_1\mathbf{S}_1)^{-1}(\mathbf{r}_1\mathbf{S}_1)^T\right]$$

At the correct location of the transmitter, $\delta\mathbf{f}_1$ becomes an eigenvector of Γ, and its corresponding eigenvalue is one. This fact may be used to determine the transmitter location through an exhaustive process of comparing the computed eigenvalue with one. Then, the corresponding eigenvector is proportional to $(\delta\mathbf{f}/f_c)$.

EXAMPLE 4.4

The sensor array shown in Figure 4.7 is split into two subarrays; the dashed line is the dividing line. The transmitter is moving along the x-axis at a constant speed of 10 m/sec. At the time of the Doppler shift measurements, the transmitter is located at (x = 10, y = 10, z = 100). Computed normalized Doppler shifts $(\delta\mathbf{f}/f_c)$ are shown in Table 4.3. Eigenvalues of Γ were computed at different values of x-coordinates, keeping the y- and z-coordinates fixed. The largest eigenvalue is always a real positive close to one and the other two eigenvalues are complex (see Table 4.4). The positive eigenvalue becomes exactly equal to one when the assumed transmitter position matches the actual transmitter location. The computed eigenvalues for different x-coordinates are shown in column two (Table 4.4). The shaded row refers to the exact position of the transmitter. The unknown scaling factor is obtained by computing the ratio (average) of the eigenvector to the Doppler shift vector. The estimated ratio is then used in Equation 4.58 to obtain an estimate of the motion vector. The x-component of the motion vector is shown in the last column (Table 4.4). The other two components were very small, particularly close to transmitter location.

The ratio is the scaling factor required for estimating the speed of the transmitter. Let $(\delta\mathbf{f}/f_c) = \gamma\mathbf{v}_1$ where $\mathbf{v}_1$ is the eigenvector corresponding to the unit eigenvalue. Using this result in Equation 4.59a we get

TABLE 4.3
Computed Doppler Shifts ($\delta f/f_c$) at Two Subarrays

Subarray1	Subarray 2
6.6010e–04	−6.5060e–04
2.9814e–03	−2.9348e–03
1.7277e–03	−2.1829e–03
4.8035e–19	2.2051e–03
1.8411e–03	−2.4733e–03
2.0912e–03	−9.7014e–04
6.0544e–04	−1.5525e–03

TABLE 4.4
The Largest + ve Eigenvalue of Γ (Rank Three Matrix) Is Equal to One Whenever the Assumed Transmitter Position Is at the Correct Location

Distance	Largest +ve Eigenvalue	Speed
0.0	0.930	0.0063
1.0	0.936	0.0063
5.0	0.964	0.0065
8.0	0.985	0.0066
10.0	1.00	0.0067
12.0	1.045	0.0066
15.0	1.037	0.0068
20.0	1.074	0.0070

Note: Variation of the eigenvalues as a function of distance is shown in the second column. The estimated speed (normalized) of the transmitter is shown in the third column. Actual speed is 0.0067.

$$\frac{\mathbf{u}_s}{c} = \gamma\left\{(\mathbf{S}_1^T\mathbf{r}_1^T\mathbf{r}_1\mathbf{S}_1)^{-1}(\mathbf{r}_1\mathbf{S}_1)^T\mathbf{v}_1\right\} \tag{4.61}$$

Thus, the normalized transmitter speed is estimated from Equation 4.61, given the scale factor.

4.4.1 Differential Doppler

It is possible that the transmitter is unstable and its center frequency fluctuates around some nominal value f_c, which is known, but not the fluctuations, ν. We can overcome this problem by comparing the Doppler shift with that at a common sensor, say, an anchor sensor. This is the differential Doppler. We rewrite Equation 4.56 as follows:

$$\delta f_1 = (f_c + \nu)\left(1 + \frac{f_c}{c}\mu_1\right) \approx \nu + \frac{f_c}{c}\mu_1 \tag{4.62a}$$

where

$$\mu_1 = \left(\frac{x_s(t) - x_1}{r_{s1}}u_x + \frac{y_s(t) - y_1}{r_{s1}}u_y + \frac{z_s(t)}{r_{s1}}u_z\right)$$

In deriving Equation 4.62a, we have assumed that, since $\mu \ll 1$ and $\nu \ll f_c$, $\nu\mu_1 \ll$ all other terms in the expansion. Compare the Doppler shift measured at any one sensor in the array with that measured at the anchor,

$$\delta f_a \approx \nu + \frac{f_c}{c}\mu_a$$

where

$$\mu_a = \left(\frac{x_s(t) - x_a}{r_{sa}} u_x + \frac{y_s(t) - y_a}{r_{sa}} u_y + \frac{z_s(t)}{r_{sa}} u_z \right)$$

and $[x_a, y_a]$ are coordinates of the anchor sensor. The differential Doppler shift with respect to the sensor at anchor node is

$$\delta f_1^{diff} = \delta f_1 - \delta f_a \approx \frac{f_c}{c}(\mu_1 - \mu_a)$$

Thus, we have been able to eliminate unknown fluctuations in the carrier frequency. We have assumed that the center frequency remains stable over a time interval equal to propagation time. For differential Doppler, in place of Equation 4.57, we have the following equation connecting the measured differential Doppler and the six unknown parameters (three transmitter locations and three motion components):

$$\delta \mathbf{f}^{diff} = \frac{f_c}{c}[\mathbf{r}\,\mathbf{S} - \mu_a]\mathbf{u}_s \tag{4.62b}$$

Solution of Equation 4.62b follows the same procedure as for solution of Equation 4.57, with matrices $\mathbf{S}_1$ and $\mathbf{S}_2$ now suitably modified.

4.5 DoA MEASUREMENTS

In this section, we shall focus on the use of (DoA) measurements for transmitter localization. To start with, we consider a set of M (M $\geq$ 2) anchor nodes, each equipped with an array for DoA measurements. Let $\theta_1, \theta_2, \ldots \theta_M$ be the error-free estimates of DoA. The anchor nodes are at known locations, $\mathbf{r}_m = (x_m, y_m)$, $m = 1,2,\ldots M$ but the transmitter is at an unknown location denoted by $\mathbf{r}_s = (x_s, y_s)$. It is easy to see that the DoAs are related to the transmitter and node coordinates:

$$\tan(\theta_m) = \frac{x_m - x_s}{y_m - y_s}, \quad m = 1,2,\cdots M \tag{4.63}$$

After rewriting the previous equation, we obtain

$$x_s - y_s \tan(\theta_m) = x_m - y_m \tan(\theta_m), \quad m = 1,2,\cdots M \tag{4.64}$$

There are M equations but two unknowns; this is a case of an over-determined system. But any two equations are enough to solve for two unknowns, for example, take anchor node one and anchor node two. The unknowns (x_s, y_s) are given by

$$x_0 = \frac{(x_1 - x_2) + (y_2 \tan(\theta_2) - y_1 \tan(\theta_1))}{\tan(\theta_2) - \tan(\theta_1)} \tan(\theta_1) + x_1 - y_1 \tan(\theta_1)$$

$$y_0 = \frac{(x_1 - x_2) + (y_2 \tan(\theta_2) - y_1 \tan(\theta_1))}{\tan(\theta_2) - \tan(\theta_1)}$$

When we have three anchor nodes we will have three possible pairs (1,2; 2,3; and 1,3) giving three possible solutions of transmitter locations (Figure 4.8). An average of three solutions may be close to the real answer. In general, if we have N anchors, we shall have N!/2! solutions (note: "!" stands for combinations). Redundancy, as seen previously, can also be utilized to reduce the effect of possible DoA estimates, which are bound to occur as we are using a finite number of uncalibrated sensors at each anchor.

4.5.1 Errors in Node Position

In the presence of random position errors of the anchor nodes, DoAs, even though error-free, will not merge into a single point representing the transmitter but within a blurred patch. The transmitter will lie somewhere inside this blurred patch. We can then try to find the least-squares estimate of the transmitter location. We have now a system of M equations (Equation 4.64):

$$\begin{aligned}
x_s - y_s \tan(\theta_1) &= x_1 - y_1 \tan(\theta_1) + \eta_1 \\
x_s - y_s \tan(\theta_2) &= x_2 - y_2 \tan(\theta_2) + \eta_2 \\
&\vdots \\
x_s - y_s \tan(\theta_M) &= x_M - y_M \tan(\theta_M) + \eta_M
\end{aligned}$$

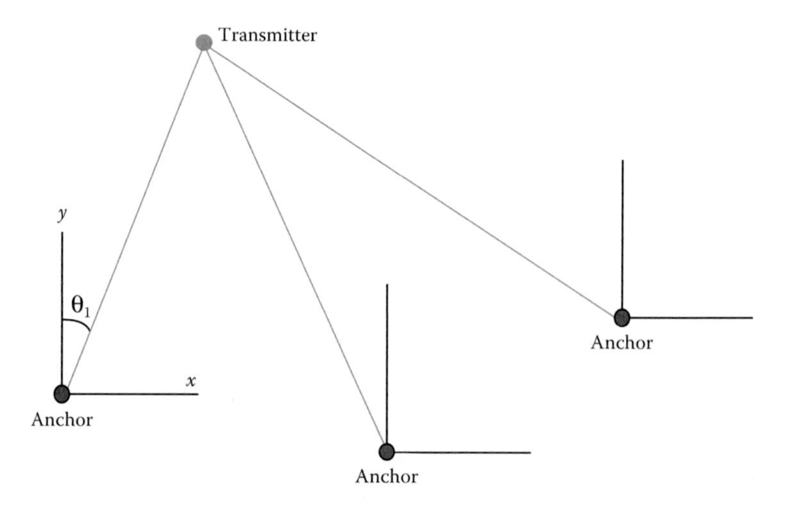

FIGURE 4.8 Three anchor nodes equipped with a sensor array for DoA estimation. In a noise-free case, all directions will point to one transmitter.

where $\eta_m = -\Delta x_m \cos(\theta_m) + \Delta y_m \sin(\theta_m)\ \Delta y_m \sin(\theta_m)$ is an error in the mth equation. Δx_m and Δy_m are errors in the position of the mth anchor node. When Δx_m and Δy_m are independent identically distributed (IID) zero mean Gaussian random variables, η_m is also a zero mean Gaussian random variable whose variance is given by

$$\text{var}\{\eta_m\} = \sigma_m^2$$

where σ_m^2 is the variance of error in the position of the mth node. We will express the previous system of equations in a matrix form as

$$\begin{bmatrix} 1 & -\tan(\theta_1) \\ 1 & -\tan(\theta_2) \\ \vdots & \\ 1 & -\tan(\theta_M) \end{bmatrix} \begin{bmatrix} x_s \\ y_s \end{bmatrix} = \begin{bmatrix} x_1 & -y_1 \tan(\theta_1) \\ x_2 & -y_2 \tan(\theta_2) \\ \vdots & \\ x_M & -y_M \tan(\theta_M) \end{bmatrix} + \begin{bmatrix} \eta_1 \\ \eta_2 \\ \vdots \\ \eta_M \end{bmatrix}$$

or

$$\mathbf{A}\mathbf{r}_s = \mathbf{b} + \boldsymbol{\eta} \tag{4.65}$$

where $\mathbf{A}$ is a system matrix, $\mathbf{b}$ is a data vector and $\boldsymbol{\eta}$ is a noise vector. The least-squares solution of Equation 4.65 is given by

$$\widehat{\mathbf{r}_0} = (\mathbf{A}^T\mathbf{A})^{-1}\mathbf{A}^T\mathbf{b} \tag{4.66}$$

It is optimal for Gaussian position errors.

4.5.2 Errors in DoA

We now assume that all DoA estimates are subject to estimation errors, presumably small compared with $\pi/2$, but node positions are error-free. The errors in DoA arise because of noise in the received signal and also due to measurement error caused by the limited resolution of a short aperture sensor array deployed at an anchor node. Let us represent erroneous the mth DoA estimate as $\hat{\theta}_m = \theta_m + \Delta\theta_m$ (see Figure 4.9). So that the angular error, $\Delta\theta_m$, may be compared with other linear measures, it needs to be mapped into a linear measure. How we do this is shown in Figure 4.9 [11]. Let $\mathbf{r}_m$ be a vector connecting the mth anchor node to the transmitter. From the transmitter location drop a perpendicular onto the measured DoA. Let $\hat{\mathbf{s}}_m$ be the vector connecting the anchor node to the base of the perpendicular, and let $\mathbf{e}_m$ be the vector representing the perpendicular. Clearly, $\mathbf{s}_m = \hat{\mathbf{s}}_m + \mathbf{e}_m$ and $\|\mathbf{e}_m\| = d_m \sin(\Delta\theta_m)$ where d_m is the distance to the transmitter from anchor node m. Let $\mathbf{r}_s$ be a vector connecting the center of the coordinates to the transmitter and $\mathbf{r}_m$ be a vector connecting the center of the coordinates to the mth anchor node. We can write that

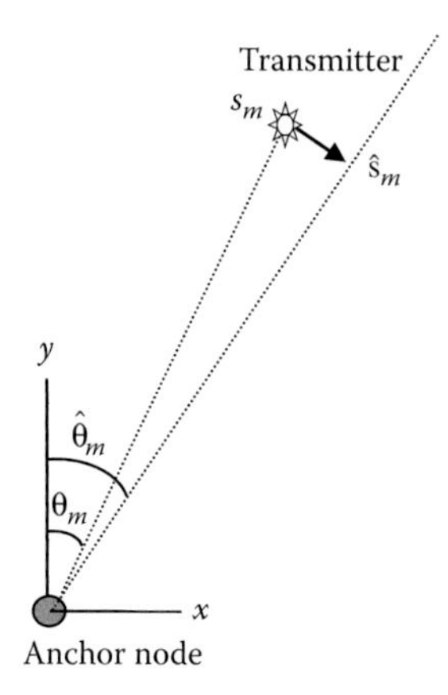

FIGURE 4.9 Mapping of an angular error $\Delta\theta_m$ $(=\hat{\theta}_m - \theta_m)$ into a linear error measure $\mathbf{e}_n$, whose magnitude is $d_m \sin(\Delta\theta_m)$. Note that no small angle approximation has been used in deriving this result.

$$\mathbf{r}_s = \mathbf{r}_m + \mathbf{s}_m \tag{4.67}$$
$$= \mathbf{r}_m + \hat{\mathbf{s}}_m + \mathbf{e}_m$$

We can eliminate $\hat{\mathbf{s}}_m$ by multiplying (dot product) with a unit vector parallel to $\mathbf{e}_m$. Let $\mathbf{a}_m$ be the unit vector in the direction of DoA and $\mathbf{a}_m^{\perp}$ be perpendicular to $\mathbf{a}_m$. In terms of DoA they are given by

$$\mathbf{a}_m = \begin{bmatrix} \sin(\hat{\theta}_m) \\ \cos(\hat{\theta}_m) \end{bmatrix}$$

and

$$\mathbf{a}^{\perp}{}_m = \begin{bmatrix} \cos(\hat{\theta}_m) \\ -\sin(\hat{\theta}_m) \end{bmatrix}$$

Multiplying both sides of Equation 4.67 with $\mathbf{a}_m^{\perp}$, we obtain

$$\mathbf{a}^{\perp}{}_m^T \mathbf{r}_s = \mathbf{a}^{\perp}{}_m^T \mathbf{r}_m + d_m \sin(\Delta\theta_m) \tag{4.68}$$

The errors in DoA may be modeled as the Gaussian random variable with a zero mean and a variance equal to σ_m^2. However, $\sin(\Delta\theta_m)$ being a non-linear transform cannot be treated as a Gaussian variable, except when $\Delta\theta_m \ll 1$ (radian). Concatenating all M equations of the Equation 4.68 type, we obtain a matrix equation

$$\begin{bmatrix} \mathbf{a}_1^{\perp^T} \\ \mathbf{a}_2^{\perp^T} \\ \vdots \\ \mathbf{a}_M^{\perp^T} \end{bmatrix} \begin{bmatrix} x_0 \\ y_0 \end{bmatrix} = \begin{bmatrix} \mathbf{a}_1^{\perp^T} \mathbf{r}_1 \\ \mathbf{a}_2^{\perp^T} \mathbf{r}_2 \\ \vdots \\ \mathbf{a}_M^{\perp^T} \mathbf{r}_M \end{bmatrix} + \begin{bmatrix} d_1 \Delta\theta_1 \\ d_2 \Delta\theta_2 \\ \vdots \\ d_M \Delta\theta_M \end{bmatrix} \tag{4.69a}$$

$$\mathbf{A}\mathbf{r}_0 = \mathbf{b} + \boldsymbol{\eta} \tag{4.69b}$$

where $\boldsymbol{\eta}$ is white the Gaussian random variable with a zero mean but variable variance. Let its covariance matrix be

$$\mathbf{W} = diag\left\{d_1^2\sigma_1^2 \; d_2^2\sigma_2^2 \cdots d_M^2\sigma_M^2 \right\}$$

We can convert non-uniform to uniform noise simply by multiplying both sides of Equation 4.69b with $\mathbf{W}^{-\frac{1}{2}}$. The least-squares solution of Equation 4.69b after conversion is

$$\hat{\mathbf{r}}_0 = (\mathbf{A}^T\mathbf{W}\mathbf{A})^{-1}\mathbf{A}^T\mathbf{W}\mathbf{b} \tag{4.70}$$

In practice, the distance from anchor nodes to the transmitter is not known. The errors in the DoA estimation may be assumed uniform when the direction-finding arrays are identical and the emitter is in the far field. Then, we may approximate $\mathbf{W} = d_0^2\sigma_0^2\mathbf{I}$ where d_0 is unknown but a constant and σ_0^2 is an unknown error variance. Equation 4.70 may be simplified to

$$\hat{\mathbf{r}}_0 = (\mathbf{A}^T\mathbf{A})^{-1}\mathbf{A}^T\mathbf{b} \tag{4.71}$$

This result is known after Stansfield [12] first derived it in 1947. The assumption of constant distance to all nodes is untenable in practice. A way out is to begin with constant distance and estimate the transmitter location. Using the estimated transmitter location, we compute distances to all nodes and recompute an improved emitter location. This step is repeated until the estimates converge to a stable value. We shall demonstrate convergence through a numerical example (Example 4.5).

EXAMPLE 4.5

Eight anchor nodes, each equipped with a linear array, are randomly distributed around the transmitter located at (red dot at 24 m, 24 m) as shown in Figure 4.10a. We assumed the position of nodes is known exactly but there is an error in the DoA estimation. The DoA errors are zero mean and have 0.01 pi^2 variance. Assuming constant distance to all nodes, as in Stansfield, we estimated the first iteration with Equation 4.71. In subsequent iterations, we used the estimated coordinates of the transmitter. As shown in Figure 4.10b, the results converge within a few (about three) iterations, though not to exact values. Evidently there is a bias in the estimation. This aspect is analyzed in the next subsection.

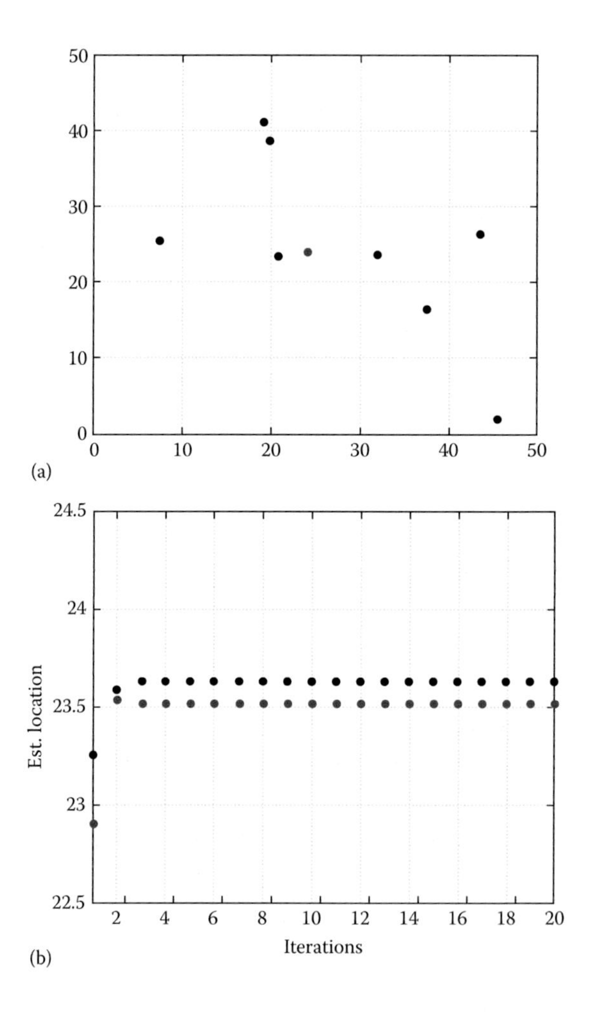

FIGURE 4.10 Estimation of the transmitter location from DoA estimation from randomly distributed nodes equipped with a linear array. The first iteration is derived assuming the transmitter is at (14 m, 14 m). Black dots stand for x_s and red dots stand for y_s.

4.5.3 Analysis of Bias

Using Equation 4.69b

$$\mathbf{A}\mathbf{r}_0 - \mathbf{b} = \boldsymbol{\eta} \tag{4.72}$$

and multiplying with $(\mathbf{A}^T\mathbf{W}^{-1}\mathbf{A})^{-1}\mathbf{A}^T\mathbf{W}^{-1}$ on both sides of the previous equation, we obtain

$$(\mathbf{A}^T\mathbf{W}^{-1}\mathbf{A})^{-1}\mathbf{A}^T\mathbf{W}^{-1}(\mathbf{A}\mathbf{r}_0 - \mathbf{b}) = (\mathbf{A}^T\mathbf{W}^{-1}\mathbf{A})^{-1}\mathbf{A}^T\mathbf{W}^{-1}\boldsymbol{\eta}$$

$$\mathbf{r}_0 - \widehat{\mathbf{r}}_0 = (\mathbf{A}^T\mathbf{W}^{-1}\mathbf{A})^{-1}\mathbf{A}^T\mathbf{W}^{-1}\boldsymbol{\eta}$$

$$\Delta \mathbf{r}_0 = (\mathbf{A}^T \mathbf{W}^{-1} \mathbf{A})^{-1} \mathbf{A}^T \mathbf{W}^{-1} \boldsymbol{\eta} \tag{4.73}$$

where $\Delta \mathbf{r}_0$ is the error in the estimate. Let us now compute its mean

$$E\{\Delta \mathbf{r}_0\} = E\left\{(\mathbf{A}^T \mathbf{W}^{-1} \mathbf{A})^{-1} \mathbf{A}^T \mathbf{W}^{-1} \boldsymbol{\eta}\right\} \tag{4.74}$$

$\boldsymbol{\eta} = [\mathrm{d}_1 \sin(\Delta\theta_1)\ d_2 \sin(\Delta\theta_2) \ldots d_N \sin(\Delta\theta_N)]^T$. Since the elements of **A** also contain $\Delta\theta_n$, $\boldsymbol{\eta}$ and **A** become correlated. $E\{\Delta \mathbf{r}_0\}$will not be equal to zero even if $E\{\boldsymbol{\eta}\} = 0$. Thus, localization estimates remain biased except for very large N (number of sensors) [13].

4.5.4 Total Least Squares (TLS)

In the least-squares method, all errors are clubbed into data vector **b**, on the right-hand side as in Equation 4.65 or in 4.69b but the system matrix **A** is also subject to DoA errors. We can improve the accuracy and bias of location estimates by invoking the concept of TLS. Formally, the TLS estimate, $\mathbf{r}_{TLS}$, is given as a solution of the following constrained optimization problem [11]:

$$\min_{(\mathbf{A}-\Lambda)\mathbf{r}_{TLS}=\mathbf{b}-\boldsymbol{\eta}} \left\| \Lambda - \boldsymbol{\eta} \right\|_F \tag{4.75}$$

where Λ is the error matrix in the system matrix, (**A**), and $\|\cdot\|_F$ stands for the Frobenius norm of a matrix. The Frobenius norm is defined as a square root of the sum of the mod square of all matrix elements. We minimally perturb the system matrix **A** and data vector **b** while the constraint in Equation 4.75 is satisfied. Let us express this constraint in a slightly different form,

$$[\mathbf{A}\ \mathbf{b}] \begin{bmatrix} \hat{\mathbf{r}}_{TLS} \\ -1 \end{bmatrix} = [\Lambda\ \boldsymbol{\eta}] \begin{bmatrix} \hat{\mathbf{r}}_{TLS} \\ -1 \end{bmatrix} \tag{4.76}$$

Pre-multiply on both sides of Equation 4.76 with their respective Hermitian transposes. After rearranging the product, we obtain

$$\begin{bmatrix} \hat{\mathbf{r}}_{TLS} \\ -1 \end{bmatrix}^H \left\{ \left([\mathbf{A}\ \mathbf{b}]^H [\mathbf{A}\ \mathbf{b}] \right) - \left([\Lambda\ \boldsymbol{\eta}]^H [\Lambda\ \boldsymbol{\eta}] \right) \right\} \begin{bmatrix} \hat{\mathbf{r}}_{TLS} \\ -1 \end{bmatrix} = 0 \tag{4.77}$$

While the rank of the entire matrix within the curly brackets (size: (3×3)) is at utmost two (for localization in two dimensions), the matrices within each of the round brackets are full rank. This rank reduction is possible only when singular value decomposition (SVD) of $[\Lambda\ \boldsymbol{\eta}]$ is a subset of SVD of [**A b**]. Further, because of Equation 4.75, we must choose the least significant singular vector of [**A b**] to create $[\Lambda\ \boldsymbol{\eta}]$. Hence, the solution of Equation 4.75 is the eigenvector of $[\mathbf{A}\ \mathbf{b}]^H[\mathbf{A}\ \mathbf{b}]$, corresponding to the least significant eigenvalue. Let $\mathbf{v}_3$ be such an eigenvector, then the solution for $\mathbf{r}_{TLS}$ will be given by

TABLE 4.5
Results of Three Independent Experiments to Estimate the Transmitter Location Using the Total Least Squares (TLS)

Experiment	x-Coordinate	y-Coordinate
1	23.899	23.398
2	24.035	24.085
3	24.458	23.851

Note: Sensor locations are in error (uniformly distributed in the range ± 2.0 m), but errors in DoA are gaussian random variables with zero mean and variance (0.0001 pi^2).

$$\widehat{\mathbf{r}}_{TLS} = -[v_{31}/v_{33} \;\; v_{32}/v_{33}]^T \tag{4.78}$$

where $\mathbf{v}_3 = [v_{31}, v_{32}, v_{33}]^T$. In order to verify the algorithm, a simple test was carried out. We used the same DSA as in Example 4.5, with errors both in DoA estimation and in sensor location. The errors in DoA are zero mean and 0.0001 pi^2 variance Gaussian random variables. The errors in sensor location are uniformly distributed over ±2 m. The results of three independent experiments are shown in Table 4.5.

4.6 COMPRESSIVE SENSING

The basic idea in compressive sensing is combining random sampling with reconstruction so that the original signal being sampled is recovered under a condition that signals are sparse, which we shall elaborate on later. Consider the following situation: We have several sinusoids lying within, say, ±100 Hz band. We would then need to sample at least at 200 samples per second, ideally over an infinite duration. If we do not know the bandwidth, we will have to first sample at a much higher rate and then compute its Fourier transform and discard all Fourier coefficients of a very small magnitude and recompute its inverse Fourier transform to reconstruct the original signal, with perhaps a small error.

Now, imagine we combine both operations even without knowing the distribution of the sinusoids over an unknown frequency band and are able to reconstruct the signal with the help of just a few samples. This, in essence, is compressive sensing. This is possible only when the signal in question is sparse, that is, it has finite degrees of freedom. For the previous example, let $\mathbf{f}$ be a signal vector having a discrete Fourier transform (DFT) representation

$$\mathbf{f} = \Psi\mathbf{c} \tag{4.79}$$

where

$$\boldsymbol{\Psi} = \text{Fourier matrix} \begin{Bmatrix} \psi_{mn} = e^{j2\pi mn/N} \\ m = 0, \cdots, N-1 \\ n = 0, \cdots, N-1 \end{Bmatrix}$$

$$\mathbf{c} = \text{Vector } (N \times 1)$$

c is a sparse vector with s large magnitude coefficients (degrees of freedom) and the remaining (N−s) coefficients are very small. The large coefficients can be anywhere within the frequency band. ψ is the representation matrix. Next, we define a sampling matrix, $\mathbf{\Phi}$ of size (M × N) where we assume that M < N (actually $M = O(s\log N)$ [1]). The elements of the sampling matrix are drawn from an IID Gaussian distribution, zero mean and 1/M variance [14; Equation 1.2]) or from the Bernoulli distribution of random variables. The observation vector (data) **y** is the product of the sampling matrix, representation matrix, and signal vector

$$\begin{aligned} \mathbf{y} &= \mathbf{\Phi\Psi c} \\ &= \mathbf{Ac} \end{aligned} \tag{4.80}$$

where $\mathbf{A} = \mathbf{\Phi\Psi}(M\text{x}N)$ is known as a sensing matrix. Since **A** can't be inverted, as its rank is ≤ M, we cannot estimate **c** from **y**. However, a least-squares (l_2 norm) solution is possible ($\mathbf{c}=(\mathbf{A}^T\ \mathbf{A})\ \mathbf{A}^T\ \mathbf{y}$ where $(.)^=$ stands for the Moore-Penrose inverse) but it is not unique. The central theme of compressive sensing is to use l_1 norm subject to a linear constraint

$$\begin{aligned} &\arg\min_{\mathbf{c}} \|\mathbf{c}\|_1 \\ &\text{subject to } \mathbf{y} = \mathbf{Ac} \end{aligned} \tag{4.81a}$$

or in presence of measurement noise of finite variance ε^2

$$\begin{aligned} &\arg\min_{\mathbf{c}} \|\mathbf{c}\|_1 \\ &\text{subject to } \|(\mathbf{y} - \mathbf{Ac})\|_2 \le \varepsilon^2 \end{aligned} \tag{4.81b}$$

and recover the sparse signal from the random observation vector (data). Further, the sensing matrix **A** must satisfy a special condition called *restricted isometry property* (RIP) [15] defined as

$$(1-\delta_S)\|\mathbf{c}\|_{l_2}^2 \le \|\mathbf{Ac}\|_{l_2}^2 \le (1+\delta_S)\|\mathbf{c}\|_{l}^2$$

where δ_s, $s = 1,2,\ldots$ is a small number, $\ll 1$ for all s-sparse vectors **c**. The RIP condition states that every set of s columns of **A** is approximately orthogonal. Further it implies that the power in **y** is close to that in **c**. The RIP condition is often satisfied by many random matrices (where matrix elements are IID random variables) and when the number of measurements $M \ge \gamma s\log(N/s)$, where γ is a constant (> 0).

The earlier optimization problem is a convex relaxation of the original *basis pursuit* algorithm involving a quasi-norm (number of non-zero elements) in place of the l_1 norm, which is supposed to promote a sparse solution, a fact that was known many decades before in reflection seismology [1].

4.6.1 Compressive Sampling Matching Pursuit (CoSaMP)

An alternate approach to sparse recovery is via an iterative algorithm, which finds the support (location of peaks) of the sparse signal iteratively. This algorithm is often known as *matching pursuit* (MP). Briefly, the basic idea in MP is as follows: Since $\mathbf{A}$ is approximately orthogonal, $\mathbf{A}^*\mathbf{A}$ shall be approximately diagonal; hence

$$\mathbf{A}^*\mathbf{y} = \mathbf{A}^*\mathbf{A}\mathbf{c} = \mathbf{c}^{proxy}$$

Thus, peaks in $\mathbf{c}^{proxy}$ are located at the same place as those of the unknown sparse signal, $\mathbf{c}$. This vital observation is used to build support of the sparse signal vector.

Using the current samples (say, in the kth iteration), we compute a signal proxy as follows: Construct submatrix, $\hat{\mathbf{A}}_k$, by picking those columns from $\mathbf{A}$ corresponding to the dominant values of the signal proxy. Let $\mathbf{A}_I$ be the union of $\hat{\mathbf{A}}_k$ and $\hat{\mathbf{A}}_{k-1}$. The least-squares estimate of unknown signal using $\mathbf{A}_I$ is

$$\hat{\mathbf{c}}_k = (\mathbf{A}_I{}^*\mathbf{A}_I)^{-1}\mathbf{A}_I{}^*\mathbf{y} \tag{4.82}$$

Determine support of $\hat{\mathbf{c}}_k$ and set all values outside this support to zero. Compute contribution of $\hat{\mathbf{c}}_k$ to the measured samples

$$\hat{\mathbf{y}}_k = \mathbf{A}_I\hat{\mathbf{c}}_k \tag{4.83}$$

Next, we compute the residue obtained by subtracting it from the measured samples, the contribution of the selected peaks until now

$$\mathbf{y}_{res} = \mathbf{y} - \hat{\mathbf{y}}_k$$

The unexplained residue $\mathbf{y}_{res}$ is now subjected to the same processing steps as previously mentioned in an iterative manner until the residue becomes very small or below the noise level. The iteration initiated with $\hat{\mathbf{y}}_0 = \mathbf{y}$, $\hat{\mathbf{c}}_0 = \mathbf{0}$, and $\hat{\mathbf{A}}_0 = \mathbf{0}$. MP has been a topic of intense research pioneered by Tropp and Needell (see [16] and [14]).

EXAMPLE 4.6

In this example, we demonstrate the power of a CoSa algorithm in comparison with a least-squares algorithm. Input consists of a sum of three-unit amplitude sinusoids (f1 = 22/128, f2 = 24/128, and f3 = 60/128) in presence of Gaussian random noise. The variance of the noise is selected, such that SNR = 20 dB. At least 64 observations were required to obtain consistent results for any type of random sampling matrix, ϕ. Notice that we have deliberately selected frequencies as rational numbers with the denominator as 128. Thus, continuous distribution is modeled by 128 discrete samples of which only 6 are active (2 for each sinusoid) and the remaining samples are equal to 0. Results are shown in Figure 4.11. While the frequencies are correctly estimated, there is a small error in estimation of amplitudes. Since the sampling matrix is randomly selected, the results can vary with the choice of the sampling matrix. For correct amplitude estimation, we need a larger SNR (> 30 dB).

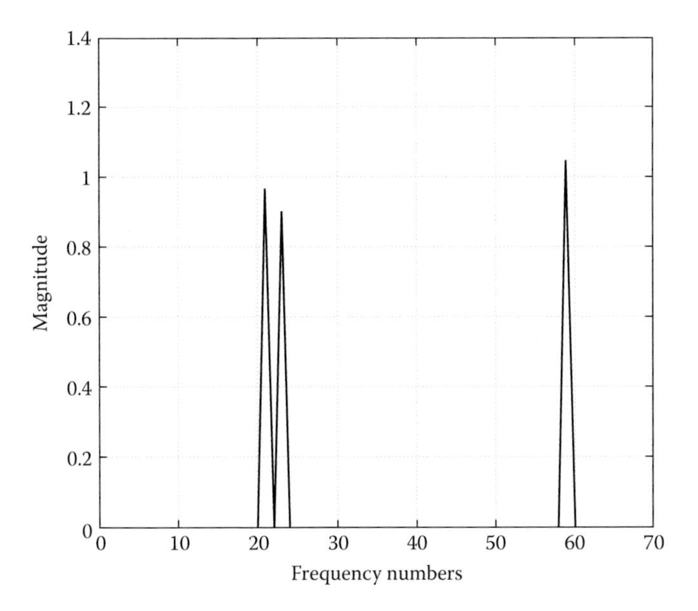

FIGURE 4.11 CoSa algorithm is used to estimate three sinusoids, the frequency and amplitude of each. The output consists of 64 random samples of 3 sinusoids in the presence of random noise (SNR = 20 dB).

For comparison, we have computed the least-squares estimate of the magnitude of sinusoids using Equation 4.80. The results are shown in Figure 4.12.

4.6.2 Localization

We shall now apply the idea of compressive sensing to source localization with DSA. It may be an acoustic source (e.g., a submarine over an array of sensors sitting on the ocean bottom) or an electromagnetic (EM) source (e.g., a radar being monitored with distributed EM sensors). We shall consider two situations, namely, (a) the transmitter is in the far field, when the wave front the incident on the array becomes planar, and (b) the transmitter is in the near field, when the wave front at the array is non-planar or curved. In the far-field case we can estimate only the DoA, but in the near-field case we can estimate both DoA and radial distance. As before, we assume that the array and the transmitter are on the same plane, although this may not be possible in reality.

4.6.3 Far Field

When a transmitter is in the far field, the only parameter of interest is the DoA. Therefore, we now consider DoA estimation using a group of distributed sensors. There are M sensors and N transmitters. In a frequency domain, the array response (vector) is given by

$$\mathbf{a}_m = \left[\gamma_{m,n} e^{j\omega_0 \frac{d_m}{c} \sin(\theta_{m,n})}, n = 0,1,\cdots,N-1 \right]$$

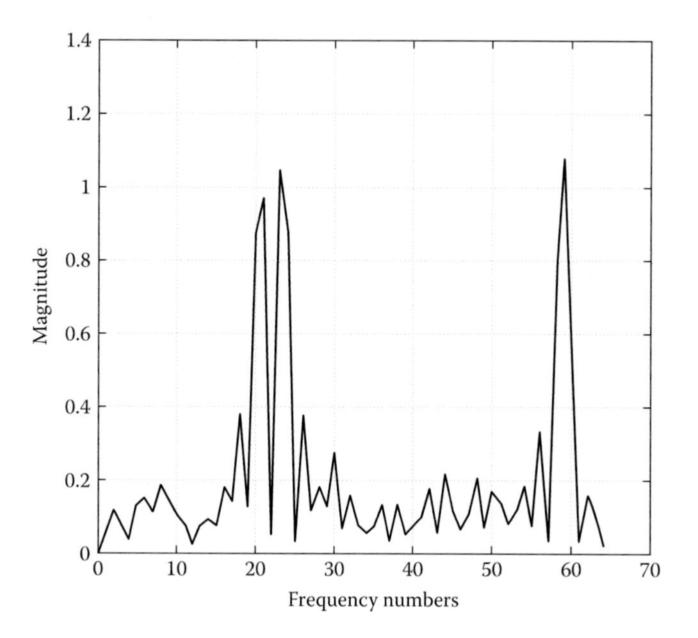

FIGURE 4.12 FFT (magnitude) output of linearly sampled 64 samples of the same signal used in Figure 4.11.

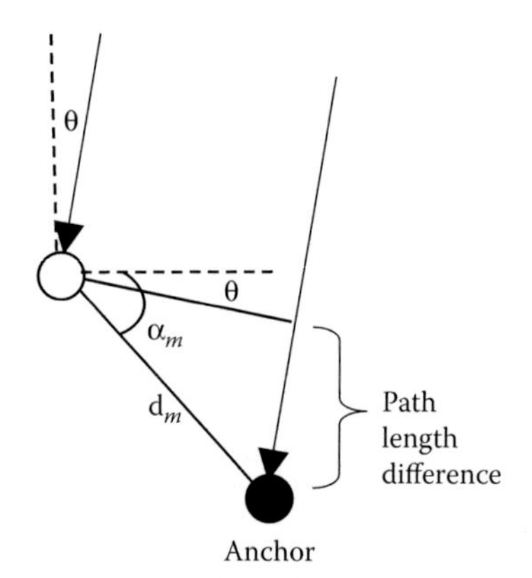

FIGURE 4.13 Path length difference relative to the anchor node. Note that the sensor at the anchor node acts as a reference sensor.

where

$$\theta_{m,n} = \alpha_m - \frac{2\pi n}{N},$$

$$\alpha_m = \tan^{-1}(\frac{y_m}{x_m}), \text{ and } d_m = \sqrt{x_m^2 + y_m^2}$$

for $m = 1, 2, \ldots M$ and γ_{mn} the mth sensor response toward the nth transmitter. DoA of nth transmitter is $2\pi n/N$ (= θ, see Figure 4.13). Note that it is assumed that the DFT processor exists at the anchor node. For an ideal sensor with an omnidirectional response, γ_{mn} is constant, but for non-ideal sensors with a random orientation, γ_{mn} is likely to be a random variable. One simple way to design a sensor with a random response is to tightly cluster together several single sensors (refer to Section 1.4 for more details).

We model the transmitter as continuously distributed over a circle of infinite radius. But in the present analysis, we shall model it as closely spaced discrete sources. Define a source vector as

$$\mathbf{x}(\boldsymbol{\omega}_0) = \left[x_n(\boldsymbol{\omega}_0),\ n = 0,1,\cdots,N-1\right]^T$$

Where $x_n(\omega_0)$ is the Fourier coefficient of the nth discrete source at an angular position $2\pi n/N$, where N is a number of equispaced discrete sources. In a limiting case as $N \rightarrow \infty$, the discrete model reduces to a continuous model. It is possible that only some of the discrete sources are active and the remaining are dormant. Thus, the source vector may be treated as a sparse vector. The array output vector is defined as

$$\mathbf{f} = \left[f_0(\boldsymbol{\omega}_0),\ f_1(\boldsymbol{\omega}_0),\cdots f_{M-1}(\boldsymbol{\omega}_0)\right]^T$$

The array output may be expressed as a matrix product of the array response matrix and the source distribution vector (signal),

$$\mathbf{f} = \begin{bmatrix} \mathbf{a}_0 \\ \mathbf{a}_1 \\ \vdots \\ \mathbf{a}_{M-1} \end{bmatrix} \mathbf{x} = \mathbf{A}\mathbf{x}$$

where $\mathbf{A}$ is an array response (or sensing matrix) matrix of size ($M \times N$, $M \ll N$). To admit a sparse solution for $\mathbf{x}$, we must show that $\mathbf{A}$ satisfies the RIP and that its columns are uncorrelated. Because the sensors are randomly distributed and not precisely calibrated, we may treat γ_{mn} as complex IID random variables. In such a situation, the rows of $\mathbf{A}$ are likely to be uncorrelated but its columns are not, unless the sources are well separated. A simple numerical experiment was designed to verify orthogonality of the $\mathbf{A}$ matrix. An array of 24 randomly distributed sensors with 2 transmitters in the far-field region was assumed. When transmitters are at 0.1 and 0.12 rad (measured with respect to the x-axis) the product $\mathbf{A}'*\mathbf{A}$ is

$$\mathbf{A}'*\mathbf{A} = \begin{bmatrix} 10.6637 - j0.0000 & -1.0929 + j0.0001 \\ -1.0929 - j0.0001 & 10.5304 - j0.0000 \end{bmatrix}$$

but when the transmitters are placed further apart at 0.1 and 0.2, (radians) the product $\mathbf{A}'*\mathbf{A}$ is

$$\mathbf{A}'*\mathbf{A} = \begin{bmatrix} 9.67729 - j0.00000 & -0.11915 + j0.00045 \\ -0.11915 - j0.00045 & 11.83259 - j0.00000 \end{bmatrix}$$

The columns become uncorrelated for larger transmitter separation. The nature of intercolumn correlation is, however, difficult to model.

EXAMPLE 4.7

This example demonstrates the power of compressive sensing for DOA estimation with randomly distributed sensors. The basic assumption is that there are known numbers of sources with power well above the background noise. DSA with 32 sensors, which are randomly distributed over an aperture of 10λ, is shown in Figure 4.14. We assumed 2 equal amplitude (10 units) sources at 10° and 13° transmitting monochromatic ($\lambda = 1$) radiation. Background noise is assumed as zero mean unit variance white complex Gaussian noise. As emphasized earlier, the columns of the gamma matrix must be uncorrelated, which would require the sources to be sufficiently apart. In this example, we have assumed three adjacent columns of the gamma matrix are correlated. This is achieved by taking the running average of three adjacent columns. The rows of the gamma matrix are, however, completely uncorrelated as sensors are randomly oriented. In Figure 4.15a, we show the result of a compressive sensing algorithm for DOA estimation in the absence of noise. The results are perfect, including estimates of the signal amplitude. In Figure 4.15b, we show the results from the previous example, rerun with added background noise.

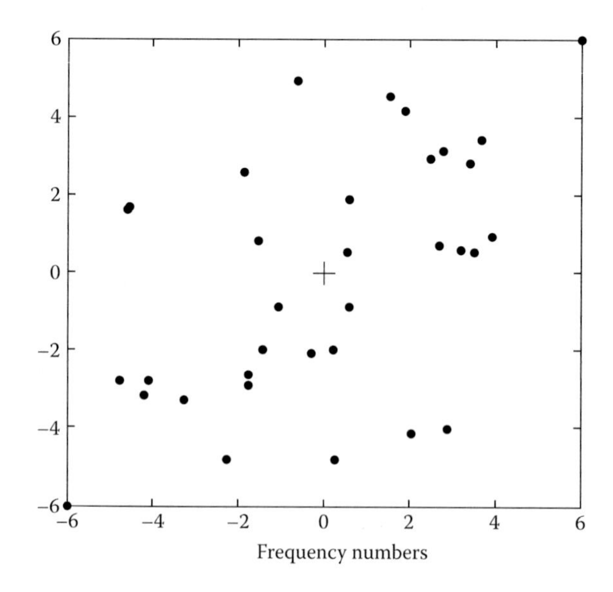

FIGURE 4.14 DSA. Sensors are randomly space over a ± 5λ rectangle. The cross denotes the anchor node.

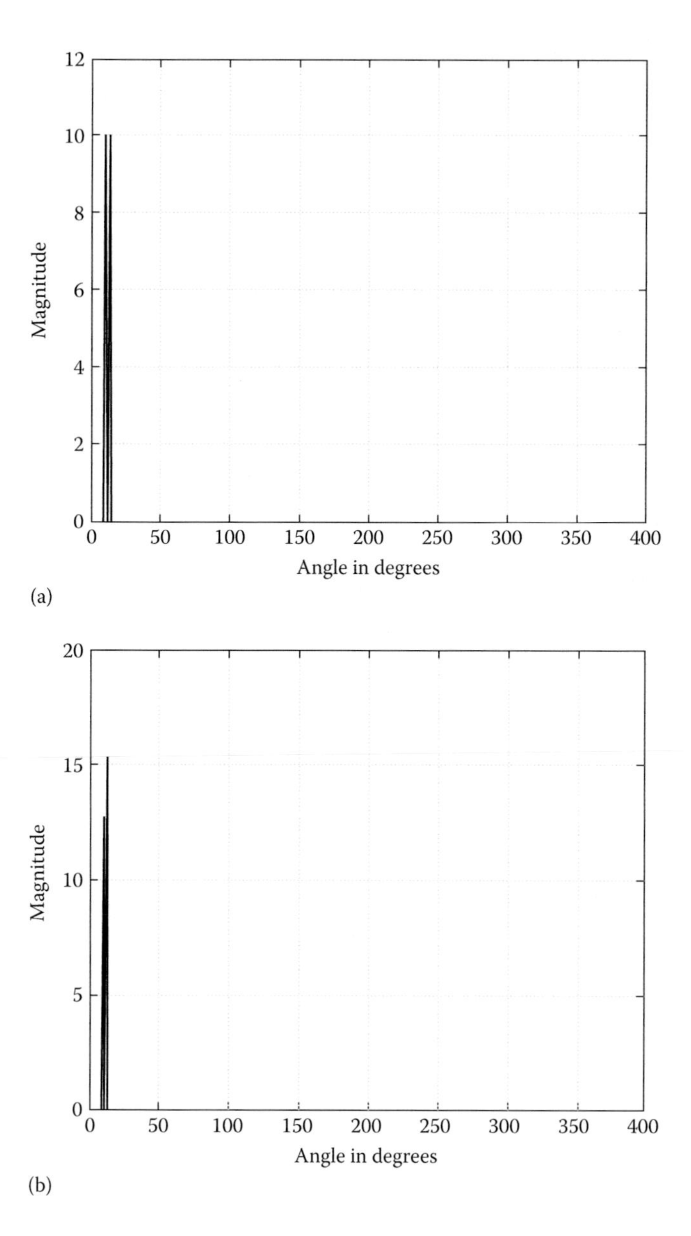

FIGURE 4.15 (a) DoA estimates with DSA having 32 elements spread over a $5\lambda \times 5\lambda$ aperture. We assumed 2 equal amplitude sources at 10° and 13° transmitting monochromatic ($\lambda = 1$) radiation. A noise-free case. The peak location and magnitude are as assumed. (b) DoA Estimates with DSA as in Figure 4.13(a) except that the array output is now contaminated with zero-mean Gaussian noise (SNR = 10). There is an error of one degree in the location of the second peak.

The DOA estimates are error-free, but estimates of signal power are incorrect, depending upon the background noise power (del = 0.5624).

4.6.4 Near Field

We now let the transmitter be at a finite distance. ToA will depend upon the transmitter location, that is, $(x_n\ y_n)$ coordinates of the transmitter. We start with an assumption that all transmitters are on a circle of radius r_n, and centered at the anchor node, in which case, we can express its coordinates in terms of a single parameter θ_n(DoA)

$$x_n = r_n\ \sin(\theta_n)\ \ y_n = r_n\ \cos(\theta_n)$$

Referring to Figure 4.16 we can compute path length difference between anchor node to source, r_n, and between the mth sensor to the source, r_m.

$$\begin{aligned}
\Delta r_{m,n} &= r_n - r_m \\
&= r_n - \sqrt{(x_m - r_n \sin(\theta_n))^2 + (y_m - r_n \cos(\theta_n))^2} \\
&= r_n \left\{ 1 - \sqrt{\left(\frac{x_m}{r_n} - \sin(\theta_n)\right)^2 + \left(\frac{y_m}{r_n} - \cos(\theta_n)\right)^2} \right\} \\
&= r_n \left\{ 1 - \sqrt{1 + \frac{d_m^2}{r_n^2} - 2\frac{x_m}{r_n}\sin(\theta_n) - 2\frac{y_m}{r_n}\cos(\theta_n)} \right\}
\end{aligned} \tag{4.84}$$

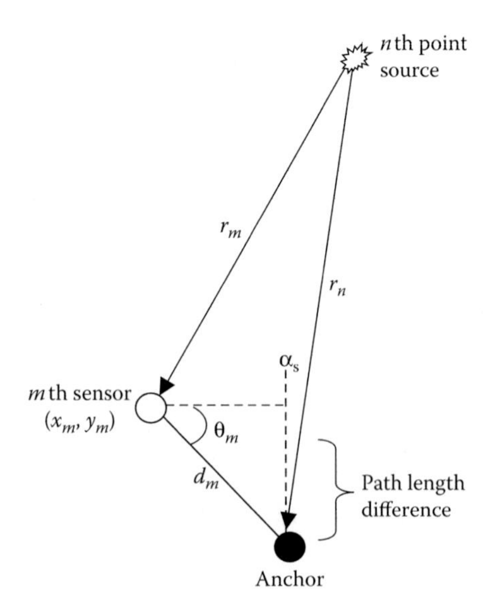

FIGURE 4.16 Computation of path length difference when the source is in the near field.

Where **A** matrix (M x N) is given by

$$\mathbf{A} = \begin{bmatrix} \gamma_{m,n} e^{j\omega_0 \frac{\Delta r_{m,n}}{c}}, & \begin{matrix} m = 0, 1, \cdots, M-1 \\ n = 0, 1, \cdots, N-1 \end{matrix} \end{bmatrix}$$

All sources are distributed over the circle of radius, r_n, centered at the anchor node. In discrete a domain, we model as point transmitters separated by an angle $2\pi/N$ where N ($\gg 1$) denotes the number of discrete transmitters used to model an otherwise a continuous transmitter. Further, we assume that only a few point transmitters are active and the rest are inactive, thus giving a sparse transmitter model. As in the near-field case, the gamma matrix is a random matrix whose rows are uncorrelated and its columns are correlated over a finite angular range. We need to estimate two parameters, namely, DoA and radial distance to the source. It turns out that the output of the compressive sensing algorithm is complex for all ranges except at the correct range and DoA. Two or more sources can be dealt with; as long as they are all at the same range

EXAMPLE 4.8

In order to demonstrate a compressive sensing algorithm, we use the parameters of DSA from the previous example. Now, there is a single source at range 10, DoA 90° and an amplitude of 12 units. The output of the algorithm for different ranges and DoAs is listed in Table 4.6. We notice that at the correct range and DoA the estimated emitter strength is real and equal to the actual source strength (in a noise-free case). This observation remains approximately valid even in presence of noise, but only in a large SNR. Further, the principle of range estimation can be extended to multiple emitters. To obtain the correct results, the emitters must be well separated, for example, a DSA with an aperture of 16λ (32 sensors), two transmitters with an angular distance of 30°, and linear distance of 15λ were correctly estimated by applying the previously mentioned principle of emitter location.

TABLE 4.6
A Source Is Assumed at DoA 90° and Range Is Equal to Ten Units

DoA ↓	Range→ 8	9	10	11	12
88^0	0.0+0.0j	0.0+0.0j	0.0+0.0j	0.0+0.0j	0.0+0.0j
89^0	0.0+0.0j	0.0+0.0j	0.0+0.0j	0.0+0.0j	0.0+0.0j
90^0	5.4695+ 4.4389j	7.5929+ 2.4628j	**8.00+0.0j**	7.2840 – 1.8700j	6.1658 – 2.9706j
91^0	0.0+0.0j	0.0+0.0j	0.0+0.0j	0.0+0.0j	0.0+0.0j
92^0	0.0+0.0j	0.0+0.0j	0.0+0.0j	0.0+0.0j	0.0+0.0j

Note: The strength of the source is eight units. The outputs for different ranges and are shown. the non-zero outputs are complex except at the correct source position shown in bold numbers.

4.6.5 One-Bit Quantized Measurements

In distributed sensor array, in order to save power, the communication overheads must be minimized. This is well achieved when each sensor output is sampled to a single bit (only sign) and yet is still able to localize a transmitter. Quantization to one bit is particularly attractive for hardware implementation in terms of inexpensive hardware, low memory space requirements, and low transmission cost. Since the one-bit (only $\pm$ 1) measurements do not provide amplitude information, the signal can be reconstructed within an unknown scaling factor. To resolve this issue, the reconstructed signal is set to a unit norm and the unknown scaling factor is determined later using other information, if available.

An important principle of consistent reconstruction is invoked for the reconstruction of the signal from one-bit measurements. Consistent reconstruction means that if the reconstructed signal is to be measured and quantized with the same system, then it should produce the same measurements as the ones at hand. In a general case, reconstruction in the presence of measurement noise using the constrained optimization, as in Equation 4.81, is consistent [17], but this is not the case if the measurement noise is due to quantization. Specifically, every noise component will have the magnitude $\leq \Delta/2$ where Δ is the uniform linear quantization step size. In this case, consistent reconstruction will produce a signal that satisfies

$$\left|(\mathbf{A}\hat{\mathbf{c}} - \mathbf{y})_i\right| \leq \frac{\Delta}{2}$$

where $(\cdot)_i$ stands for the ith component of the vector. Now, consistent reconstruction is not possible. Instead, what is expected is

$$sign(\mathbf{a}_1 \cdot \hat{\mathbf{c}}) = y_i \quad \forall \quad i$$

or alternatively,

$$y_i(\mathbf{a}_i \cdot \hat{\mathbf{c}}) \geq 0 \tag{4.85a}$$

In a compact matrix form, the previous result of Equation 4.85a) may be expressed as

$$diag\{\mathbf{y}\}\mathbf{A}\hat{\mathbf{c}} \geq \mathbf{0} \tag{4.85b}$$

For the reconstruction of CoSa with sign output (one bit), we have an additional constraint, Equation 4.85b in addition to the minimum norm one. Thus, it is necessary to solve the following optimization problem:

$$\begin{aligned} &\arg\min \|\mathbf{c}\|_1 \\ &\text{subject to } diag\{\mathbf{y}\}\mathbf{A}\mathbf{c} \geq \mathbf{0} \\ &\|\mathbf{c}\|_2 = 1 \end{aligned} \tag{4.86}$$

Since the minimization is performed on a unit sphere, the problem is not convex, and hence convergence to a global minimum is not guaranteed [17]. Another algorithm, inspired by CoSaMP, is the MSP, which does iterative greedy search to compute the sparse minimum. Specifically, MSP updates a sparse estimate of the signal $\hat{\mathbf{c}}$ by iterating the following procedure until convergence [18]. Iteration begins with the initial estimate, $\hat{\mathbf{c}}^0 = \mathbf{0}$. Define a signal proxy as

$$\mathbf{c}^{prox} = \mathbf{A}^T\mathbf{y}$$

1. Identify which sign constraints are violated. Define a vector, $\mathbf{r}^k = (\mathbf{c}^{prox})^-$, where $(\cdot)^-$ stands for the negative part of each element of the vector (constraint violators). All positive elements are set to zero.
2. Update signal proxy in the kth iteration as $\mathbf{c}^{prox} = \mathbf{A}^T diag(\mathbf{y})\mathbf{r}^k$.
3. Form a union of support $\mathbf{c}^{prox}$ by selecting 2K components with the largest magnitude with the support of $\hat{\mathbf{c}}^{k-1}$. That is, $T^k = \text{supp}(\mathbf{c}^{prox}) \cup \text{supp}(\hat{\mathbf{c}}^{k-1})$.
4. Perform consistent reconstruction over $T^k$by minimizing $\left\|(diag(\mathbf{y})\mathbf{Ac})^-\right\|_2^2$ subject to $\left\|\mathbf{c}\right\|^2 = 1$ and $\mathbf{c} = 0$ outside T^k.
5. Update signal estimate: $\mathbf{c}^k$

Further details, including an Octave program, are given in [2,18]

EXAMPLE 4.9

In this example, we consider a harmonic analysis problem taken from Example 4.6, but with an amplitude quantized to one bit (that is, ± 1). The length of the measured data (after quantization) is equal to the signal length (N = m = 128). The SNR before quantization was set to 10 dB. We have used both one-bit CoSa and a direct fast Fourier transform (FFT) approach. The results are shown in Figure 4.17a,b. The frequencies are correctly identified in the CoSa approach but in the FFT approach they are not. In fact, the output as seen in Figure 4.17b is totally erratic.

A DSA, having a large number of scattered sensors but with limited power, is an ideal case for one-bit compressive sensing. First, we explore the prospect of estimating the DoA of single and multiple narrowband point sources in a far-field scenario. A plane wavefront is incident on the array resulting in relative time delays or phase delays as measurable information. Given the sensor locations, we can relate the transmitter DoA to the relative phase difference, Equation 4.84. The sensors are assumed not calibrated and randomly oriented. Yet, the columns of the sensing matrix may not be fully uncorrelated. This was taken into account, as outlined previously. Sensor output is sampled and quantized to two levels (± 1). We now need to transmit just one bit/sample, which represents a huge saving on communication bandwidth and power. But this saving comes at the cost of very large measured data. The array is required to be a few thousand sensors. All sensors must be time synchronized. A computer simulation experiment to measure the DoA of two closely spaced point sources with help of a 2000-sensor array is given next.

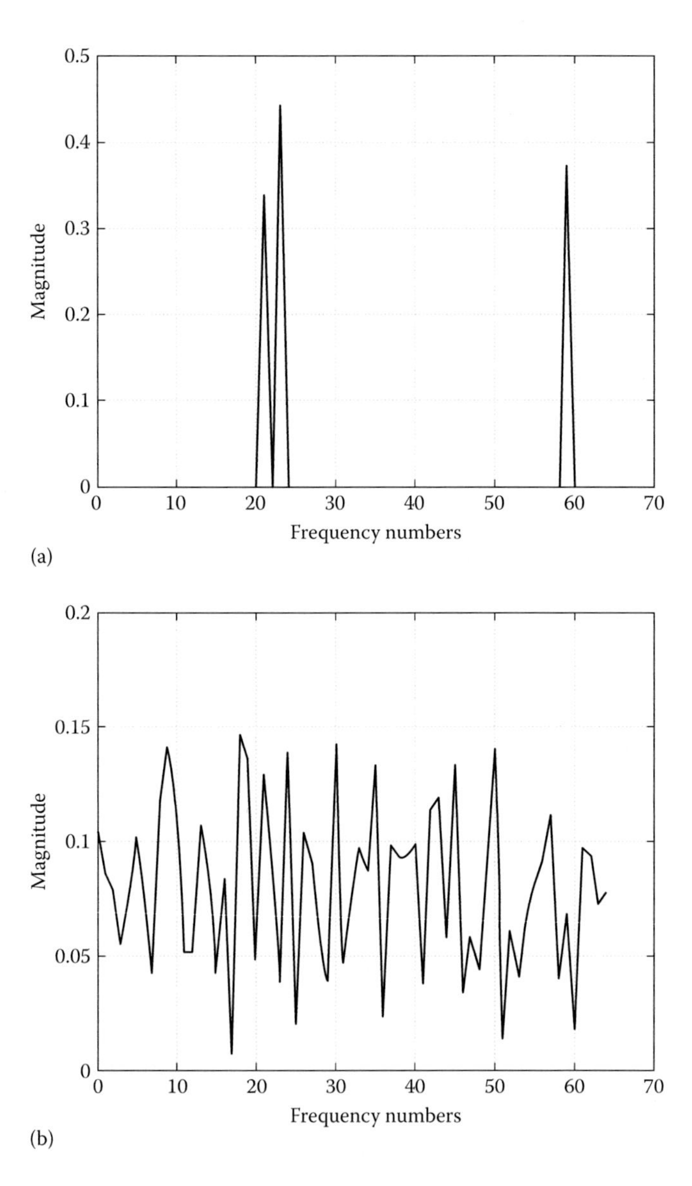

FIGURE 4.17 Harmonic analysis of a signal containing three sinusoids plus white Gaussian noise (snr = 10 dB). The signal was quantized to one bit (± 1). Result from the CoSa approach is shown in (a), and that from the FFT approach in (b).

EXAMPLE 4.10

A sensor array consisting of 2000 randomly placed and randomly oriented sensors is assumed. The array aperture is 10λ with the anchor node at its center. Each sensor can directly communicate with the anchor, where processing hardware (including a DFT processor) and software are located. A DSA over a rectangular patch ($10 \times 10\lambda^2$) is shown in Figure 4.18a. Two point sources in the far

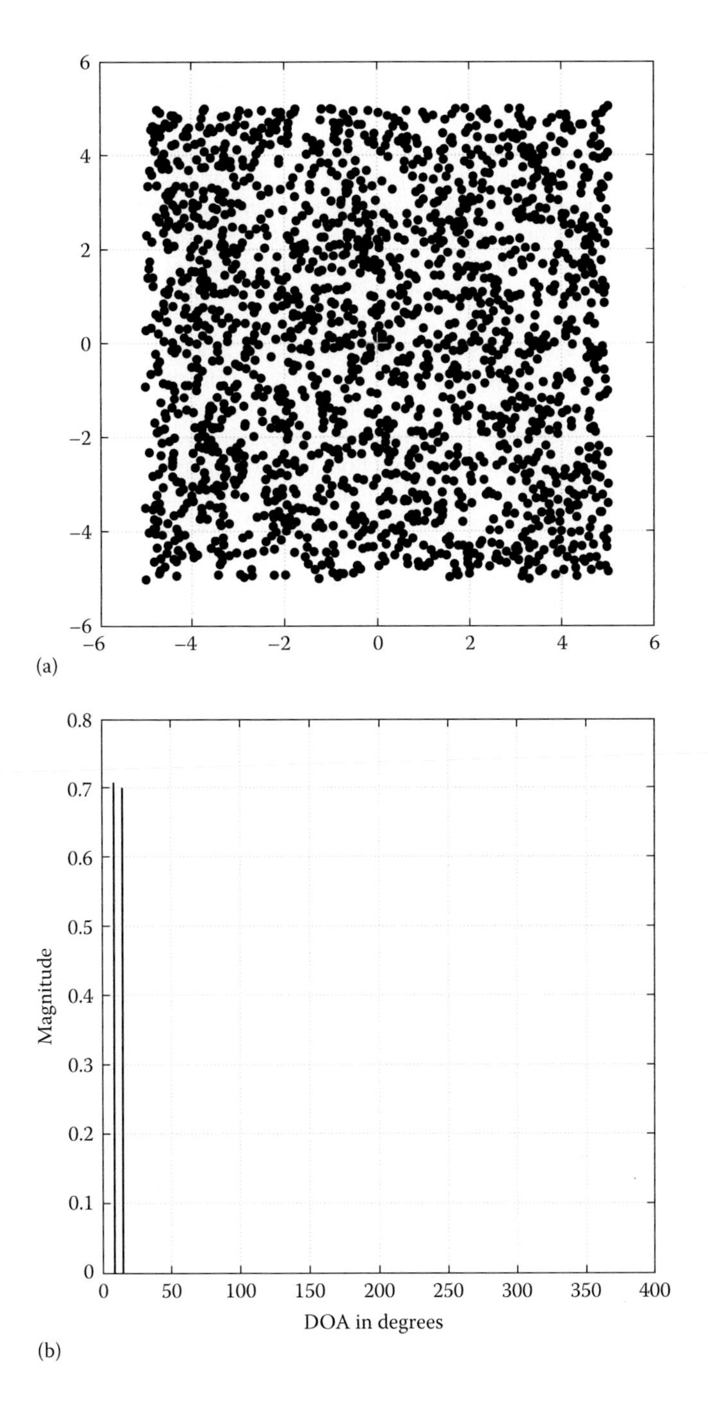

FIGURE 4.18 (a) Each dot represents a sensor. There are 2000 sensors with an anchor node at the center, shown as a red cross (b) Result of CoSa one-bit estimation. The two peaks shown are at correct DoAs, 9° and 15° respectively. Background noise in the sensor output (before quantization) is 10 dB.

field (the wavefront is planar over the array aperture) radiate narrowband signals with the same center frequency. The sources may be correlated, as if one is the reflected signal of the other. The DoAs are 9° and 15° respectively. The output of each sensor is corrupted with Gaussian white noise with such a variance, such that SNR = 10 dB. The noisy signal is now sampled and quantized to two levels (± 1). The final result of the CoSa algorithm applied to one-bit data is shown in Figure 4.18b. Peaks appear at correct DoAs. The estimate of amplitude of each source also appears to be in line with the amplitude of the radiated signal, which for mathematical convenience is treated as complex, but for practical reasons we consider only the real part and hence the half power.

The same numerical experiment was carried out to localize a source at a finite range. The wave front is now curved. We need to estimate the difference in ToA from the transmitter at a sensor with reference to the anchor node. Derivation of a sensing matrix in terms of the transmitter location turns out somewhat involved, which we had derived earlier in this section. We have assumed a no-noise situation in order to demonstrate the principle of range estimation, namely, the transmitter's estimated strength *at correct range must be real*. This is a practical requirement. This property, as shown in Table 4.7, is proved for one specific emitter location but is found to be true for all locations outside the array aperture. The estimated strength becomes complex at all other assumed ranges. Interestingly the DoA estimation is practically independent of assumed range. This is useful a property as DoA is first estimated with only an approximate knowledge of the range. While the result shown in Table 4.7 is for a no-noise case, the numerical experiments seem to indicate that the principle of range estimation holds good even at as low an SNR as zero dB.

A favorable localization result, obtained with one-bit quantization, is possible only at the cost of a large number of measurements, in our case a large number of sensors. Experimentally, it is found that the imaginary part of the estimated transmitter strength goes to zero asymptotically, as the number of sensors goes to infinity (tested up to 4000 sensors). However, we obtained consistent results for a 1000 or above sensor array. For less than 1000 sensors, for example, for 100 sensors, instances of a few failures (the minimum imaginary part not being at the correct location) were encountered.

TABLE 4.7
Estimated Transmitter Strengths Using One-Bit Input Data for Different DoAs and Ranges

DOA ↓	**Range→8**	**9**	**10**	**11**	**12**
118°	0.0+0.0j	0.0+0.0j	0.0+0.0j	0.0+0.0j	0.0+0.0j
119°	0.0+0.0j	0.0+0.0j	0.0+0.0j	0.0+0.0j	0.0+0.0j
120°	0.8318–0.5550j	0.9639–0.2661j	**0.9996+0.0268j**	0 .9650+0.2620	0 .9031+0.4293j
121°	0.0+0.0j	0.0+0.0j	0.0+0.0j	0.0+0.0j	0.0+0.0j
122°	0.0+0.0j	0.0+0.0j	0.0+0.0j	0.0+0.0j	0.0+0.0j

Note: The uncalibrated sensors (2000 sensors) are randomly placed and oriented as shown in Figure 4.16a. a single point pmitter is assumed at range 10λ and direction 120°.

4.7 SUMMARY

The received signal strength is the simplest parameter a sensor can measure without any need for synchronization or calibration. Only the signal strength ratio between a pair of sensors is relevant in the estimation of location coordinates. If one of the sensors is a common reference sensor, the unknown transmitter location may be obtained as a least-squares solution of system linear equations. Measurement errors do introduce errors in the estimates, but imposing the constraint that the location coordinates and range are not independent can marginally reduce errors.

ToA requires clocks both at the transmitter and the sensor to be precisely synchronized. In the absence of noise, basic ToA equations can be linearized and solved easily, but in the presence of noise linearization is not possible, as we cannot remove quadratic terms involving the transmitter location. An iterative method along with distance constraints has been proposed. Convergence is quite rapid.

TDoA works without a cooperative transmitter but we need to synchronize all sensors. Straightforward minimization of observed time differences minus computed TDoAs for the assumed transmitter position leads to a highly non-linear minimization problem, which may not have a unique solution. Several alternate approaches have been suggested. In one method, which we explain at length, the right-hand term is linearized but at the cost of introducing one unknown term in the left-hand side. First, we solve for the linear quantities in a closed form in terms of the unknown term. Minimization of this expression results into a quadratic equation, which is easily solved. This approach has been termed as the minimization of "equation error." The minimization of observed time difference minus computed TDoAs is termed as *output error minimization*. The two approaches may lead to different answers. Other approaches involve considering the unknown term on the left-hand side as unknown but linear; so, it may be combined with other unknowns and solved jointly. In yet another method, devising an annihilating matrix cleverly eliminates the unknown term on the left-hand side.

FDoA arises whenever the transmitter or sensors are in motion, which causes a Doppler shift in frequency. Since the shift is dependent on the velocity of the moving transmitter (or sensor) and its current location, we can estimate both from the observed shift information. For this, DSA is split into two non-overlapping subgroups, each having three or more sensors. It is found that at the correct transmitter location, the maximum eigenvalue of a certain matrix is one and the corresponding eigenvector determines the velocity vector. We can overcome the problem of a fluctuating center frequency by computing the differential Doppler with reference to an anchor.

The basic idea in compressive sensing is to combine random sampling with reconstruction so that the original signal being sampled is recovered under a condition that signals are sparse. The central theme of compressive sensing is to use the l_1 norm subject to a linear constraint. An alternate to convex relaxation is an iterative algorithm, which finds the support (location of peaks) of the sparse signal iteratively. This algorithm is often known as MP. We have applied this concept to the localization of a transmitter either in the far field or near field. A large number of discrete transmitters are placed over an infinite circle (the far-field case), but most transmitters are silent. Only a few are active making the input signal sparse. Further, we assume that the sensors are directional but randomly oriented. High-quality DoA estimates were obtained. The method worked very well, even with one-bit quantization.

REFERENCES

1. E. J. Candes and M. Wakin, An introduction to compressive sampling, *IEEE Signal Processing Magazine*, pp. 21–30, 2008.
2. L. Jaques, J. N. Laska, P.T. Boufounos, and R. G. Baranuik, Robust 1-bit compressive sensing via binary stable embeddings of sparse vectors, *IEEE Transactions on Information Theory*, vol. 59, pp. 2082–2102, 2013.
3. D. Li and Y. H. Hu, Energy-based collaborative source localization using acoustic microsensor array, *EURASIP Journal on Applied Signal Processing*, pp. 321–337, 2003.
4. J. J. Caffery, A new approach to the geometry of TOA location, *IEEE*, VTC2000, pp. 1943–1949.
5. Y. T. Chan, C. H. Yau, and P. C. Ching, Linear and approximate maximum likelihood localization from TOA measurements, *IEEE 2003, Signal Processing and Its Applications. Proceedings*, vol. 2, pp. 295–298, 2003
6. J. O. Smith and J. S. Abel, Closed-form least-squares source location estimation from range-difference measurements, *IEEE Transactions on Speech and Signal Processing*, vol. ASSP-35, pp. 1661–1669, 1987
7. Y. T. Chan and K. C. Ho, An efficient closed-form localization solution from time difference of arrival measurements, *IEEE* ICASSP-94, pp. II-393–396, 1994,
8. B. Freidlander, A passive localization algorithm and its accuracy analysis, *IEEE Journal of Ocean Engineering*, vol. OE-12, pp. 234–245, 1987.
9. W. R. Hahn and S. A. Tretter, Optimum processing for delay-vector in passive signal arrays, *Transactions on Information Theory*, vol. IT-19, pp. 608–614, 1973.
10. K. Yao, R. E. Hudson, C. W. Reed, D. Chen, and F. Lorenzelli, Blind beamforming on a randomly distributed sensor array system, *IEEE Journal on Selected Areas in Communications*, vol. 16, pp. 1555–1567, 1998.
11. K. Dogancay, Bearins-only target localization using total least squares, *Signal Processing*, vol. 85, pp. 1695–1710, 2005.
12. R. Stansfield, Statistical theory of d.f. fixing, *Journal of the Institute of Electrical Engineers*, vol. 94, pp. 762–770, 1947.
13. K. Dogancay, Bias compensation for the bearings-only pseudo linear target track estimator, *IEEE Transactions on Signal Processing*, vol. 54, pp. 59–68, 2006
14. D. Needell and J. A. Tropp, CoSaMP: Iterative signal recovery from incomplete and inaccurate samples, *Applied and Computational Harmonic Analysis*, vol. 26, pp. 301–321, 2009.
15. R. Baranuik, M. Davenport, R. De Vore, and M. Wakin, *A Simple Proof of the Restricted Isometry Property for Random Matrices, Constructive Approximation*, New York, NY: Springer, 2008.
16. J. A. Tropp and A. Gilbert, Signal recovery from random measurements via orthogonal matching pursuit, *IEEE Transactions on Information Theory*, vol. 53, pp. 4655–4666, 2007.
17. P. T. Boufounos and R. G. Baraniuk, 1-bit compressive sensing, *Proceedings of the 42nd Conference on Informational Sciences and Systems (CISS)*, Princeton, NJ, 2008.
18. P. T. Boufounos and R. G. Baraniuk, Reconstructing sparse signals from their zero crossings, *IEEE, ICASSP*, pp. 3361–3364, 2008.
19. Y. Tomita, A. A. H. Damen, and P. M. J. Van Den Hof: Equation error versus output error methods, *Ergonomics*, vol. 35, pp. 551–564, 1992.

5 Hybrid Methods for Localization

Until now, we have concentrated on the use of distance information, which is derived from the time of arrival (ToA)/time difference of arrival (TDoA) measurements and the direction of arrival (DoA) information, which is derived through a local regular sensor array (e.g., linear array) in place of a single sensor or through multi-component sensors (e.g., a three-component acoustic array). It is hoped that the combined use of all parameters, ToA, TDoA, frequency difference of arrival (FDoA), and DoA, or a subset of them will yield a more efficient localization algorithm, for example, as shown in the next section, where we show how a non-linear term can be removed in combined use of FDoA and TDoA. Here, we explore this line of approach.

5.1 COMBINED USE OF FDoA AND TDoA

Localization of a source becomes a simple linear problem with the combined use of Doppler shift information (FDoA) and TDoA.

5.1.1 Narrow Band

First we rewrite Equation 2.9d in a different form:

$$
\begin{aligned}
r_{s1}\delta f_1 &= \frac{f_c}{c}\left((x_s(t)-x_1)u_x+(y_s(t)-y_1)u_y+z_s(t)u_z\right)\\
r_{s2}\delta f_2 &= \frac{f_c}{c}\left((x_s(t)-x_2)u_x+(y_s(t)-y_2)u_y+z_s(t)u_z\right)\\
&\cdots\\
r_{sm}\delta f_m &= \frac{f_c}{c}\left((x_s(t)-x_m)u_x+(y_s(t)-y_m)u_y+z_s(t)u_z\right)
\end{aligned}
\tag{5.1}
$$

where u_x, u_y, u_z are three velocity components of a moving transmitter, and $x_s(t)$, $y_s(t)$, and $z_s(t)$ are the coordinates of a transmitter at time t. r_{sm} is the range of the transmitter from the mth sensor. Subtract the mth equation from the first equation and obtain, after some simplification

$$
\begin{aligned}
(r_{sm}\delta f_m - r_{s1}\delta f_1) &= r_{s1}(\delta f_m-\delta f_1)+(r_{sm}-r_{s1})\delta f_m \\
&= \frac{f_c}{c}\left[(x_1-x_m)u_x+(y_1-y_m)u_y\right]
\end{aligned}
\tag{5.2}
$$

Notice that in Equation 5.2 we have used both FDoA as well as TDoA. Note that $r_{sm} - r_{s1} = c\tau_{m1}$, where τ_{m1} is TDoA between the first and the mth sensor. We shall treat r_{s1} along with u_x and u_y as unknowns. We have the following linear equation:

$$(r_{sm} - r_{s1})\delta f_m = \frac{f_c}{c}\left[(x_1 - x_m)u_x + (y_1 - y_m)u_y\right] - r_{s1}(\delta f_m - \delta f_1)$$

There will be M–1 such linear equations, which we express in matrix form. Define the following vectors and matrix:

$$\mathbf{B} = \left[(r_{s2} - r_{s1})\delta f_2 \ (r_{s3} - r_{s1})\delta f_3 \cdots (r_{sM} - r_{s1})\delta f_M\right]^T$$

$$\mathbf{A} = \begin{bmatrix} \frac{f_c}{c}(x_1 - x_2) & \frac{f_c}{c}(y_1 - y_2) & -(\delta f_2 - \delta f_1) \\ \frac{f_c}{c}(x_1 - x_3) & \frac{f_c}{c}(y_1 - y_3) & -(\delta f_3 - \delta f_1) \\ \vdots & & \\ \frac{f_c}{c}(x_1 - x_M) & \frac{f_c}{c}(y_1 - y_M) & -(\delta f_M - \delta f_1) \end{bmatrix}$$

$$\boldsymbol{\theta} = \left[u_x \ u_y \ r_{s1}\right]^T$$

$$\mathbf{B} = \mathbf{A}\boldsymbol{\theta} \tag{5.3}$$

The least-squares solution of θ is given by

$$\boldsymbol{\theta} = \left(\mathbf{A}^T\mathbf{A}\right)^{-1}\mathbf{A}^T\mathbf{B} \tag{5.4}$$

Note that vector **B** is known from the observed FDoA and TDoA. From Equation 5.4, we obtain two x and y components of transmitter motion and the distance to the transmitter from the reference sensor, numbered as one. We select another sensor, call it two, and estimate the distance to the transmitter using the method outlined earlier. Expressing the square of the distances explicitly, we get two equations:

$$r_{s1}^2 = (x_s - x_1)^2 + (y_s - y_1)^2 + z_s^2 \tag{5.5a}$$

$$r_{s2}^2 = (x_s - x_2)^2 + (y_s - y_2)^2 + z_s^2 \tag{5.5b}$$

We subtract Equation 5.5a from 5.5b and obtain

$$\frac{1}{2}(r_{s2}^2 - r_{s1}^2) = (x_1 - x_2)x_s + (y_1 - y_2)y_s \tag{5.6a}$$

Next, we select yet another reference sensor, call it three, and obtain yet another linear equation

$$\frac{1}{2}(r_{s3}^2 - r_{s1}^2) = (x_1 - x_3)x_s + (y_1 - y_3)y_s \tag{5.6b}$$

By solving Equation 5.6a and 5.6b, we obtain (x_s, y_s)

$$x_s = \frac{1}{2}\frac{(r_{s2}^2 - r_{s1}^2)(y_1 - y_3) - (r_{s3}^2 - r_{s1}^2)(y_1 - y_2)}{(x_1 - x_2)((y_1 - y_3) - (x_1 - x_3)((y_1 - y_2)} \tag{5.7a}$$

$$y_s = \frac{1}{2}\frac{(r_{s2}^2 - r_{s1}^2)(x_1 - x_3) - (r_{s3}^2 - r_{s1}^2)(x_1 - x_2)}{(x_1 - x_3)((y_1 - y_2) - (x_1 - x_2)((y_1 - y_3)}$$

Finally, we obtain the z-coordinate from Equation 5.5a

$$z_s = \sqrt{r_{s1}^2 - (x_s - x_1)^2 - (y_s - y_1)^2} \tag{5.7b}$$

Note that we have chosen a +ve sign since the transmitter is assumed to be above the sensor array.

What remains to be estimated is the vertical component of transmitter motion. For this, we go back to Equation 5.1b and express the vertical component of transmitter motion in terms of quantities already estimated:

$$u_z = \frac{1}{z_s}\left[r_{s1}\delta f_1 \frac{c}{f_c} - \left((x_s - x_1)u_x + (y_s - y_1)u_y\right)\right] \tag{5.8}$$

We need at least four sensors to estimate six parameters, namely, three source position coordinates and three velocity components. Note that Equations 5.7b and 5.8 are dependent on (x_s, y_s, u_y, u_y). In the presence of noise, however, we need many more sensors in order to get a robust estimate.

5.1.2 Annihilating Matrix

In Chapter 4, we have, in the context of position estimation from TDoA measurements, introduced the concept of an annihilating matrix to remove a non-linear term $r_{s1}\mathbf{d}$ in the main Equation 4.35a, which, for convenience, is reproduced here:

$$\mathbf{S}\hat{\mathbf{x}}_s = \frac{1}{2}(\boldsymbol{\delta} - 2\hat{r}_{s1}\mathbf{d} - 2\mathbf{T}) \tag{4.35a}$$

Taking a time derivative on both sides of the previous equation, we obtain an equation connecting transmitter motion and range rate, which in turn relates to the Doppler shift (see Section 2.5).

$$\mathbf{S}\dot{\hat{\mathbf{x}}}_s = \frac{1}{2}(\dot{\boldsymbol{\delta}} - 2\dot{\hat{r}}_{s1}\mathbf{d} - 2\hat{r}_{s1}\dot{\mathbf{d}}) \tag{5.9}$$

$$= \frac{1}{2}\dot{\boldsymbol{\delta}} - \begin{bmatrix}\mathbf{d} & \dot{\mathbf{d}}\end{bmatrix}\begin{bmatrix}\dot{\hat{r}}_{s1} \\ \hat{r}_{s1}\end{bmatrix}$$

$$\begin{bmatrix} \mathbf{S} & \\ & \mathbf{S} \end{bmatrix} \begin{bmatrix} \hat{\mathbf{x}}_s \\ \dot{\hat{\mathbf{x}}}_s \end{bmatrix} + \begin{bmatrix} \mathbf{d} & \mathbf{0} \\ \mathbf{d} & \dot{\mathbf{d}} \end{bmatrix} \begin{bmatrix} \dot{\hat{r}}_{s1} \\ \hat{r}_{s1} \end{bmatrix} = \frac{1}{2} \begin{bmatrix} \boldsymbol{\delta} - 2\mathbf{T} \\ \dot{\boldsymbol{\delta}} \end{bmatrix} \tag{5.10}$$

All vectors and matrices are defined in Section 4.3. Define an annihilating matrix **P**, which will null the second term on the left-hand side, where

$$\mathbf{H} = \begin{bmatrix} \mathbf{d} & \mathbf{0} \\ \mathbf{d} & \dot{\mathbf{d}} \end{bmatrix}$$

is a matrix of size $2(M-1)\times 2$. The annihilating matrix is actually an orthogonal projection matrix of $\mathbf{H}$,

$$\mathbf{P} = \mathbf{I} - \mathbf{H}(\mathbf{H}^T\mathbf{H})^{-1}\mathbf{H}^T \tag{5.11a}$$

By pre-multiplying on both sides of Equation 5.10 with the annihilating matrix, we obtain

$$\mathbf{P}\begin{bmatrix} \mathbf{S} & \\ & \mathbf{S} \end{bmatrix} \begin{bmatrix} \hat{\mathbf{x}}_s \\ \dot{\hat{\mathbf{x}}}_s \end{bmatrix} = \frac{1}{2}\mathbf{P} \begin{bmatrix} \boldsymbol{\delta} - 2\mathbf{T} \\ \dot{\boldsymbol{\delta}} \end{bmatrix} \tag{5.11b}$$

We are thus able to eliminate from Equation 5.10 a term involving range and range rate. These two parameters contain the unknown transmitter position and velocity in a non-linear form [1]. The least-squares solution of Equation 5.11 is straightforward, given by

$$\begin{bmatrix} \hat{\mathbf{x}}_s \\ \dot{\hat{\mathbf{x}}}_s \end{bmatrix} = \frac{1}{2} \left[\begin{bmatrix} \mathbf{S} & \\ & \mathbf{S} \end{bmatrix}^T \mathbf{P} \begin{bmatrix} \mathbf{S} & \\ & \mathbf{S} \end{bmatrix} \right]^{-1} \begin{bmatrix} \mathbf{S} & \\ & \mathbf{S} \end{bmatrix}^T \mathbf{P} \begin{bmatrix} \boldsymbol{\delta} - 2\mathbf{T} \\ \dot{\boldsymbol{\delta}} \end{bmatrix} \tag{5.12}$$

It may be noted that though the size of the annihilating matrix is $2\,(M-1)\times 2(M-1)$, the required matrix inversion is for a (4×4) matrix (for planar array but 6×6 for volume array). Issues of computational complexity and small error analysis are discussed in [1].

5.2 DIRECT POSITION DETERMINATION

The position and velocity of a transmitter have been estimated from ToA, TDoA, DoA, and FDoA information, which were independently estimated earlier from the observed data. In that approach, we have ignored the fact that there is a single transmitter that is the source for all the previously measured parameters. They are indeed not independent in this sense. Weiss [2] recognized this fact and has come up with a method wherein the position and velocity are estimated directly from the observed data in a single step. He and other authors [3–5] have shown that direct position determination (DPD) leads to more accurate estimates. In this section, we consider a non-stationary transmitter and a stationary distributed sensor array (DSA). The reverse case, where the transmitter is fixed (on the ground) but the DSA is placed on a moving platform (i.e., airplane), will not be considered here, but it is discussed in [6].

We now consider a case of a transmitter transmitting a narrow band signal with a known center frequency f_c and bandwidth Δf. The Doppler shift due to the transmitter motion is much smaller than the bandwidth (Doppler shift$\ll\Delta f$). The transmitter is moving with a constant speed and is being tracked by a cluster of stationery DSA sensors. There are M sensors, one transmitter and N time samples. The discrete signal at the mth sensor from a transmitter is represented as

$$\begin{gathered} s_m(n) = s_0(n-\tau_m)\exp(-j2\pi f_m n) + \eta_m(n) \\ \mathrm{n} = 0 \quad \cdots \quad \mathrm{N}-1 \end{gathered} \tag{5.13}$$

where τ_m is travel time from the transmitter to the mth sensor and f_m is the Doppler shift at the mth sensor due to the motion of the transmitter. η_m is noise at the mth sensor (Equation 4.56).

$$f_m = \frac{f_c}{c} \frac{\mathbf{u}^T(\mathbf{r}_0 - \mathbf{r}_m)}{\|(\mathbf{r}_0 - \mathbf{r}_m)\|}$$

where $\mathbf{u} = [u_x \ u_y]^T$ is the velocity vector and $\mathbf{r}_0 = [x_0 \ y_0 \ z_0]^T$ is the position vector of the transmitter and $\mathbf{r}_m = [x_m \ y_m \ 0]^T$ is the position vector of the mth sensor. For simplicity, we shall assume that all sensors lie in a plane ($z = 0$) and the transmitter motion is confined to a plane at height $z_p = \text{constant} > 0$. We go on to the frequency domain by evaluating the finite discrete Fourier transform (FDFT) [7] on both sides of Equation 5.13

$$\tilde{s}_m(f_k) = \tilde{s}_0(f_k - f_m)\exp(-j2\pi\tau_m f_k) + \tilde{\eta}_m(f_k) \tag{5.14}$$

where an overhead wiggle denotes the discrete Fourier transform (DFT) and $f_k = k/N,\ k = 0, \pm 1, \pm 2, \cdots \pm N/2$. To express Equation 5.14 in a matrix form, we introduce the following vectors (of length = N) and matrices (size = N × N):

$$\tilde{\mathbf{s}}_m = \left[\tilde{s}_m(f_{-N/2}) \quad \cdots \quad \tilde{s}_m(f_{N/2})\right]^T$$

$$\tilde{\boldsymbol{\eta}}_m = \left[\tilde{\eta}_m(f_{-N/2}) \quad \cdots \quad \tilde{\eta}_m(f_{N/2})\right]^T$$

$$\tilde{\mathbf{s}}_0 = \left[\tilde{s}_0(f_{-N/2}) \quad \cdots \quad \tilde{s}_0(f_{N/2})\right]^T$$

$$\mathbf{A}_m = diag\left\{e^{-j2\pi\tau_m f_{-N/2}} \quad \cdots \quad e^{-j2\pi\tau_m f_{N/2}}\right\}$$

A matrix equivalent of Equation 5.14 is given by

$$\underset{N\times 1}{\tilde{\mathbf{s}}_m = \mathbf{A}_m \mathbf{F}_m \tilde{\mathbf{s}}_0 + \tilde{\boldsymbol{\eta}}_m} \tag{5.15}$$

where $\mathbf{F}_m$ is a cycle shift operator (matrix) that is used to down/up shift vector $\tilde{\mathbf{s}}_0$ by $[Nf_m]$, where $[x]$ stands for integral part of x. For example, the required cyclic shift operator to down shift by two units a vector of a length of five (N = 5) is

$$\mathbf{F}_m = \begin{bmatrix} 0\,0\,0\,1\,0 \\ 0\,0\,0\,0\,1 \\ 1\,0\,0\,0\,0 \\ 0\,1\,0\,0\,0 \\ 0\,0\,1\,0\,0 \end{bmatrix}$$

Next, we compute a cross-spectral matrix of an array output vector defined in Equation 5.15. We assume that the signals emitted by different transmitters and their background noises are uncorrelated.

$$\begin{aligned} \mathbf{C}_{m,m'} &= E\left\{\tilde{\mathbf{s}}_m \tilde{\mathbf{s}}_{m'}^H\right\} \\ &= \mathbf{A}_m \mathbf{F}_m E\left\{\tilde{\mathbf{s}}_0 \tilde{\mathbf{s}}_0^H\right\} \mathbf{F}_{m'}^H \mathbf{A}_{m'}^H + E\left\{\tilde{\boldsymbol{\eta}}_m \tilde{\boldsymbol{\eta}}_{m'}^H\right\} \end{aligned} \tag{5.16}$$

We further assume that the background noises are temporally and spatially uncorrelated. This results in a simplification of Equation 5.16.

$$\mathbf{C}_{m,m'} = \mathbf{A}_m \mathbf{F}_m \Lambda\, \mathbf{F}_{m'}^H \mathbf{A}_{m'}^H + \sigma_0^2 \mathbf{I} \delta_{m,m'} \tag{5.17}$$

where $\Lambda = E\left\{\tilde{\mathbf{s}}_0 \tilde{\mathbf{s}}_0^H\right\}$ is a diagonal matrix (size: $N \times N$) whose diagonal elements are a spectrum of the transmitted signal. Λ is a full rank matrix, when there are no zero elements in the range 0 to N−1 discrete frequencies. It is independent of sensor location, that is, independent of m. Let σ_0^2 be variance of the background noise, also independent of the sensor location. Then, we can collect all terms dependent on m or m' into a matrix, $\mathbf{B}$ (size: $MN \times N$),

$$\mathbf{B} = \begin{bmatrix} \mathbf{A}_1 \mathbf{F}_1 \\ \vdots \\ \mathbf{A}_M \mathbf{F}_M \end{bmatrix}$$

and rewrite Equation 5.17 in terms of **B**

$$\mathbf{C} = \mathbf{B}\boldsymbol{\Lambda}\mathbf{B}^H + \sigma_0^2\mathbf{I} \tag{5.18}$$

where **C** is a complete cross-spectral matrix (size: MN × MN). As shown subsequently, the **B** matrix has an interesting property of orthogonality:

$$\begin{aligned}
\mathbf{B}^H\mathbf{B} &= \left[\mathbf{F}_1^H\mathbf{A}_1^H \cdots \mathbf{F}_M^H\mathbf{A}_M^H\right]\begin{bmatrix}\mathbf{A}_1\mathbf{F}_1 \\ \vdots \\ \mathbf{A}_M\mathbf{F}_M\end{bmatrix} \\
&= \left[\mathbf{F}_1^H\mathbf{A}_1^H\mathbf{A}_1\mathbf{F}_1 + \quad \cdots \quad + \mathbf{F}_M^H\mathbf{A}_M^H\mathbf{A}_M\mathbf{F}_M\right] \\
&= \left[\mathbf{F}_1^H\mathbf{F}_1 + \quad \cdots \quad + \mathbf{F}_M^H\mathbf{F}_M\right] \\
&= M\mathbf{I}
\end{aligned} \tag{5.19}$$

The rank of the cross-spectral matrix is determined by the rank of $\boldsymbol{\Lambda}$ or by the number of non-zero spectral values in the range $\pm\ 1/\Delta t$, where Δt is the sampling interval.

Let $\mathbf{E}_s$ represent the eigenvectors corresponding to non-zero eigenvalues of the cross-spectral matrix **C**, whose rank is N. Referring to Equation 5.18, the matrix **B** may be treated as the basis vectors of $(\mathbf{C} - \sigma_0^2\mathbf{I})$, that is, they span the signal subspace of $(\mathbf{C} - \sigma_0^2\mathbf{I})$. The signal eigenvectors span the same signal space as matrix **B**. Thus, $\mathbf{E}_s$ and **B** are linearly related, for example,

$$\mathbf{E}_s = \mathbf{B}\boldsymbol{\Phi} \tag{5.20}$$

where $\Phi N \times N$ is an orthogonal matrix

$$\boldsymbol{\Phi}^H\boldsymbol{\Phi} = \frac{1}{M}\mathbf{I}$$

Consider the matrix product $\mathbf{E}_s{}^H\tilde{\mathbf{B}}$ where $\tilde{\mathbf{B}}$ is another basis vector pertaining to the assumed transmitter parameters (location and speed). It is a correlation product between vectors spanning signal subspace $\mathbf{E}_s$ and $\tilde{\mathbf{B}}$. We shall compute the magnitude square of this correlation product and we perform a sum of all its diagonal elements. Using Equation 5.20 in the following matrix product we obtain

$$\mathbf{E}_s{}^H\tilde{\mathbf{B}} = \boldsymbol{\Phi}^H\mathbf{B}^H\tilde{\mathbf{B}} \tag{5.21}$$

or its magnitude is given by $\mathbf{E}_s{}^H\tilde{\mathbf{B}}\tilde{\mathbf{B}}^H\mathbf{E}_s$. Whenever the choice of transmitter parameters agrees with the actual, but unknown, transmitter parameters, that is, $\tilde{\mathbf{B}} = \mathbf{B}$, the magnitude product reduces to

$$\mathbf{E}_s^{\,H}\tilde{\mathbf{B}}\tilde{\mathbf{B}}^H\mathbf{E}_s = M\mathbf{I} \tag{5.22}$$

Alternatively, we can take the average of all diagonal elements and consider a simpler scalar quantity to verify the correlation between $\tilde{\mathbf{B}}$ and $\mathbf{B}$

$$\text{Corr1: } sum\left[diag\left\{\mathbf{E}_s^{\,H}\tilde{\mathbf{B}}\tilde{\mathbf{B}}^H\mathbf{E}_s\right\}\right]/N \tag{5.23}$$

whose maximum value is M (M: number of sensors).

We can also define another correlation quantity between noise space $\mathbf{E}_n$ and $\tilde{\mathbf{B}}$,

$$\text{Corr2: } \quad sum\left[diag\left\{\mathbf{E}_n^{\,H}\tilde{\mathbf{B}}\tilde{\mathbf{B}}^H\mathbf{E}_n\right\}\right]/N \tag{5.24}$$

where $\mathbf{E}_n$ are vectors spanning the noise subspace corresponding to a zero eigenvalue (or small eigenvalues in the presence of noise), Corr2 will be zero whenever $\tilde{\mathbf{B}} = \mathbf{B}$, otherwise it is non-zero. Note that the inverse of Corr2 is analogous to the MUSIC spectrum [7]. Corrl (or Corr2) will have to be computed over a grid around prior knowledge of the transmitter location and speed (a minimum of four-dimensional space, two space and two motion parameters). Much of this computation is however straightforward and it may be implemented in parallel. In the example that follows, we demonstrate the effectiveness of the algorithm using an example that is close to a real-life situation.

EXAMPLE 5.1

As an illustration of the previously mentioned algorithm, we consider a fixed 32-randomly distributed sensor array shown in Figure 5.1. The sensors are spread over an area of ± 100 m. The transmitter is located at (30, 20, 1000 m), that is, 1000 m above the plane of array. A gray dot in Figure 5.1 represents the transmitter. The speed of the moving transmitter is 100, 10, 0 m/s (x, y, and z components, respectively). The transmitter is in a class of a supersonic missile. Assume that the transmitter radiates a hopping frequency narrowband signal (bandwidth = 0.1 MHz) with a carrier frequency of 1 GHz. An example of transmitted signal and its spectrum is shown in Figure 5.2. There are 10 random sinusoids inside the signal band.

The carrier frequency is first filtered out, and the leftover baseband is then sampled at the Nyquist rate (sampling interval: 10 μs). To compute the DFT, 1600 time samples (0.0016 s long) were used. First, 16 frequency samples were used to compute a cross-spectral matrix, which is then averaged over 40 transmissions.

The rank of the cross-spectral matrix is equal to the number of DFT coefficients N used in the computation of the matrix. The number of transmissions should be greater than N. In this example, we have used 40 transmissions (N = 16). Further, the time duration of the signal used in the evaluation of DFT coefficients is long enough to make the coefficients mutually uncorrelated. This will ensure matrix Λ in Equation 5.17 is diagonal. We have used signal duration about 100 times N, that

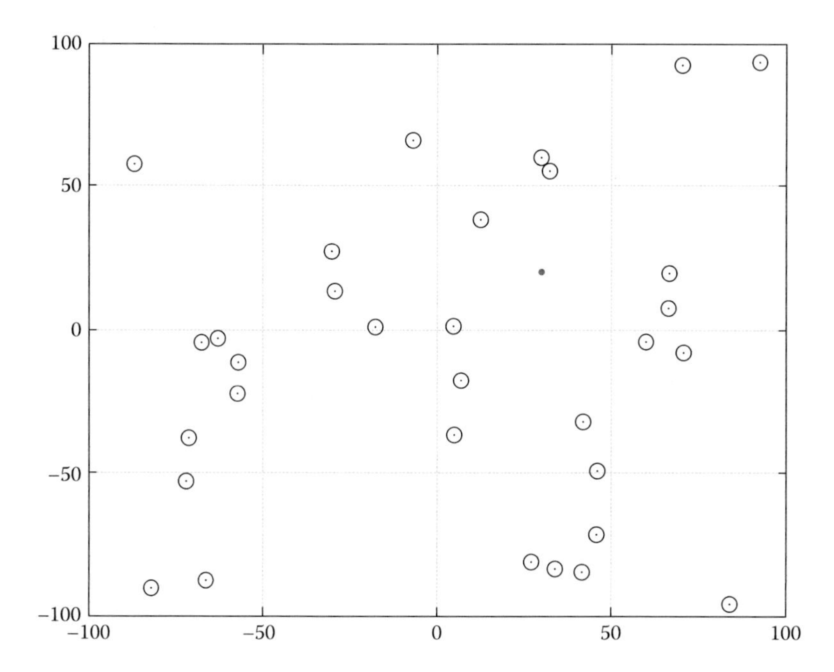

FIGURE 5.1 Thirty-two randomly distributed sensor arrays spread over ± 100 m. The transmitter is located at the gray dot 1000 m above the array.

is, 1600 time samples. At each sensor, we inject Gaussian random noise with a zero mean and variance consistent with assumed a signal-to-noise ratio (SNR) (in this case SNR = 20 on linear scale).

The cross-spectral matrix is next subjected to eigenvalue decomposition. The signal subspace is identified by large eigenvalues and the noise subspace by small eigenvalues. To compute two correlations defined in Equations 5.23 and 5.24, Corr1 and Corr2, we compute the inner product of signal/noise subspace with computed basis vectors for assumed transmitter parameters. The results are shown in Table 5.1. Each table shows scan results of one parameter keeping others fixed and known. The maximum (M = 32) of Corr1 and the minimum (zero) of Corr2 are at the correct value of the unknown parameter. It was noticed that the results showed some dependence on the actual distribution of sensors. Data length played a significant role in being able to measure the Doppler effect due to the transmitter motion. It is noticed that a data length of 20 or more times N is required for localization but a length over 70 times N would be required for velocity estimation (see Figure 5.3).

5.3 DoA FROM DISTRIBUTED SENSORS

We now consider P stationary transmitters and M sensors. Each transmitter transmits an unknown deterministic waveform, $s_p(t)$. Sensors form clusters around each anchor. All sensors within a cluster are able to communicate with an anchor node, as shown in Figure 5.4, where all signal processing is carried out.

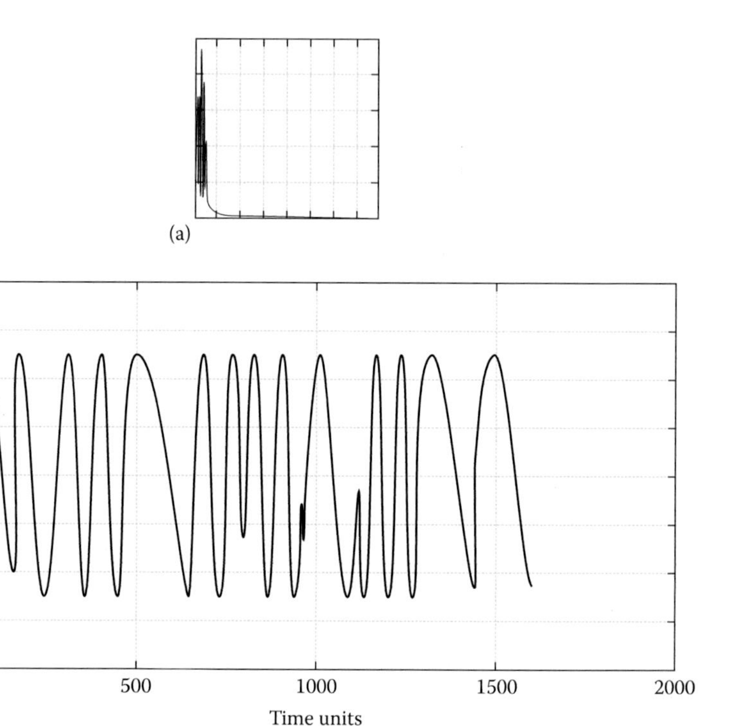

FIGURE 5.2 (a) Hopping frequency signal and (b) its spectrum. To get a normalized frequency, multiply by 7.8125×10^{-4}.

Consider an oblique triangle ABC with a pth transmitter, an mth sensor, and an anchor node as the three corners of the triangle. It is easy to show that angles α,β, and γ (shown in Figure 5.5) are as follows:

$$\alpha=\theta_m^p-\theta_0^p$$

$$\beta=180+\phi_m-\theta_m^p$$

$$\gamma=\theta_0^p-\phi_m$$

where θ_0^p is the direction to the pth transmitter measured at the anchor node, θ_m^p is the direction to the pth transmitter measured at the mth sensor, and ϕ_m is the direction to the mth sensor measured at the anchor node. All directions are with respect to the x-axis. The sine law for a triangle states that

$$\frac{r_m}{\sin(\theta_m^p-\theta_0^p)}=\frac{r_0^p}{\sin(\theta_m^p-\phi_m)}=\frac{r_m^p}{\sin(\theta_0^p-\phi_m)}, \qquad (5.26)$$

$$m=1,2\cdots M$$

TABLE 5.1

Scanning with Respect to x- or y-Coordinate Keeping Other Parameters Fixed and Known

(a) x-Component (m/s)	Corr1	Corr2
70	15	0.5357
80	15.8	0.5132
90	26.5	0.1710
100	32	1.8e-07
110	26.5	0.1771
120	21.9	0.3205
130	21.9	0.3205
(b) x-Coordinate (m)	**Corr1**	**Corr2**
0	16.2	0.5164
10	19	0.4117
20	25	0.2212
30	32	1.8e-07
40	20	0.3928
50	17	0.4968
60	16.5	0.5088
(c) y-Coordinate (m)	**Corr1**	**Corr2**
−10	28.2	0.1220
0	28.2	0.1220
10	28.2	0.1220
20	32	1.8e-07
30	30	0.0631
40	30	0.0631
50	30	0.0631

Note: In (a) Scanning in x-direction is shown and in (b) Scanning in y-direction is shown.

where r_m, r_0^p, and r_m^p are distances from the anchor node to the mth sensor, r_0^p is the distance from the anchor to the pth transmitter, and r_m^p is the distance from the mth sensor to the pth transmitter, as shown in Figure 5.5. These are shown in Figure 5.4 in terms of distances; after dividing by the speed of wave, we obtain the travel times. Equation 5.26 may be expressed in terms of familiar quantities: DoAs, θ_0^p, and θ_m^p at two sensors, namely, the anchor and the mth sensor.

$$\tau_0^p = \tau_m \frac{\sin(\theta_m^p - \phi_m)}{\sin(\theta_m^p - \theta_0^p)} \tag{5.27}$$

$$\tau_m^p = \tau_m \frac{\sin(\theta_0^p - \phi_m)}{\sin(\theta_m^p - \theta_0^p)}$$

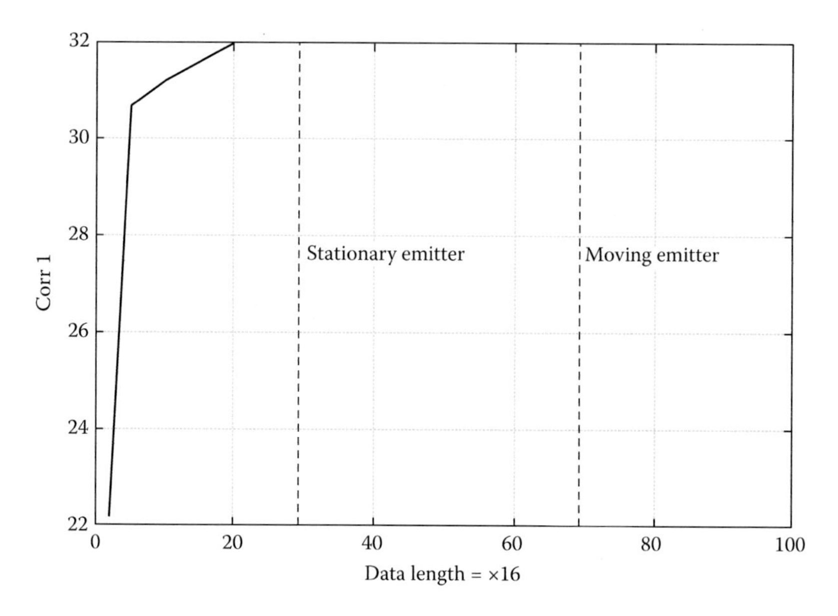

FIGURE 5.3 Correlation (corrl) as a function of data length. For a stationary transmitter, the maximum is achieved for as low as 390 samples. But for a moving transmitter, one would need at least 1120 samples.

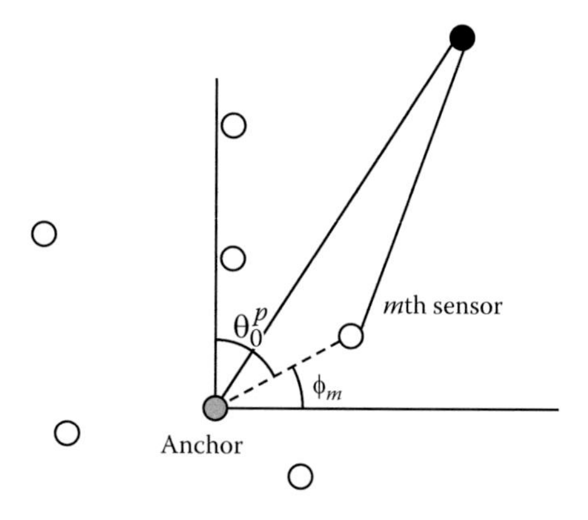

FIGURE 5.4 A cluster of sensors around an anchor within their communication range works as an array for DoA estimation. All sensors and transmitters are in the same plane.

The DoA at two sensors are thus related, though non-linearly, to the ToA. Note that $\tau_m = r_m / c$; travel time from the anchor to the mth sensor is a known quantity. Equation 5.27 provides a link between DoA and ToA.

The FDFT of N output samples is computed as

$$S_m(k) = \sum_{p=1}^{P} w_m^p \exp\left(-j\frac{2\pi}{N\Delta t} k\tau_m^p\right) S^p(k) + \mathrm{N}_m(k) \tag{5.28}$$

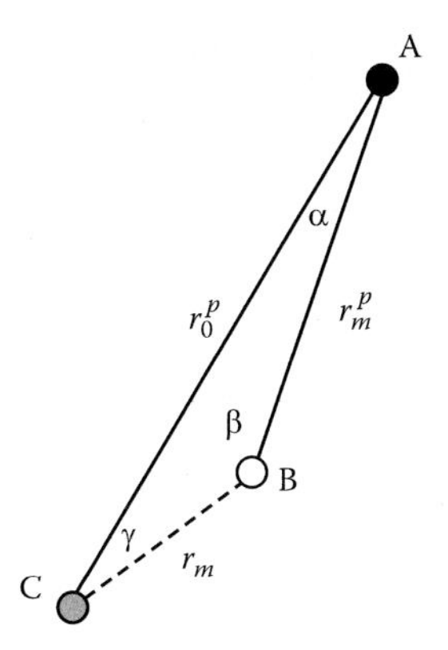

FIGURE 5.5 The *p*th transmitter (a), the *m*th sensor (b), and anchor node (c). They form a three-cornered oblique triangle, ABC.

where $S_m(k)$ is the kth FDFT coefficient of the mth sensor output. Similarly, $N_m(k)$ is the kth FDFT coefficient of noise at the mth sensor, and w_m^p is sensitivity of mth sensor towards pth transmitter.

The DoA from the pth transmitter to the mth sensor is related to τ_m^p through Equation 5.27. There will be MP (number of sensors × number of transmitters) unknowns, but in a far-field problem with the source being quite far (compared with the size of array), there are only P DoAs to the P transmitter. In Figure 5.5, let the transmitter (A) be in far field and (C) be an anchor point. The delay for the signal to reach sensor m relative to the anchor point is given by

$$\tau_m^p = \frac{r_m}{c}\cos\left(\theta_0^p - \phi_m\right) \tag{5.29}$$

Indeed, τ_m^p is TDoA between the anchor and the mth sensor from the pth transmitter. Using Equation 5.29 in Equation 5.28, we can express Equation 5.28 in a matrix form as

$$\begin{aligned}\mathbf{\Sigma} &= \left[\mathbf{d}^1 \quad \mathbf{d}^2 \quad \cdots \quad \mathbf{d}^P\right]\mathbf{S} + \mathbf{N} \\ &= \mathbf{DS} + \mathbf{N}\end{aligned} \tag{5.30}$$

where

$$\mathbf{\Sigma} = \left[S_1(k)\, S_2(k) \cdots S_M(k)\right]^T$$

$$\mathbf{d}^p = \frac{1}{\sqrt{M}}\begin{bmatrix} w_1^p \exp\left(-j\frac{2\pi}{N\Delta t}k\tau_1^p\right) & w_2^p \exp\left(-j\frac{2\pi}{N\Delta t}k\tau_2^p\right) \\ \cdots \quad w_M^p \exp\left(-j\frac{2\pi}{N\Delta t}k\tau_M^p\right) & \end{bmatrix}^T$$

$$p = 1, 2, \cdots P$$

$$\mathbf{D} = \left[\mathbf{d}^1 \quad \mathbf{d}^2 \quad \cdots \quad \mathbf{d}^P\right]_{M\times P}$$

$$\mathbf{S} = \left[S^1(k)\, S^2(k) \cdots S^P(k)\ \right]^T$$

and

$$\mathbf{N} = \left[\eta_1(k)\, \eta_2(k) \cdots \eta_M(k)\right]^T$$

We would like to obtain the least-squares estimate of **S** in the presence of non-uniform noise, that is, the noise level varies from sensor to sensor, but it is white Gaussian noise [8]. A simple method is to convert the non-uniform noise to uniform noise and then apply the least squares. This is simply done by multiplying on both sides of Equation 5.30 with the inverse of noise covariance matrix, $\boldsymbol{\Theta}^{-1}$. The least-squares estimate of **S** is given by

$$\hat{\mathbf{S}} = \left(\boldsymbol{\Theta}^{-1}\mathbf{D}\right)^{\dagger} \boldsymbol{\Theta}^{-1}\boldsymbol{\Sigma} = (\mathbf{D}')^{\dagger}\, \boldsymbol{\Sigma}' \tag{5.31}$$

where $(\mathbf{D}')^{\dagger} = \left(\mathbf{D}'^T\mathbf{D}'\right)^{-1} \mathbf{D}'^T$ is the pseudo-inverse. Primes on $\mathbf{D}'$ and Σ' denote normalization with respect to noise variance, for example, $\mathbf{D}' = \boldsymbol{\Theta}^{-1}\mathbf{D}$. The unknown parameters in signal model Equation 5.31 are

$$\mathbf{S} = \left[S^1(k)\, S^2(k) \cdots S^P(k)\right]^T \quad \text{P}\, {}^{N}\!/_{2} \text{ unknowns}$$

$$\mathbf{w}^P = \left[w_1^P\ w_2^P \cdots w_M^P\right] \quad \text{MP unknowns}$$

$$\boldsymbol{\theta}_m = \left[\tau_m^1\ \tau_m^2 \cdots \tau_m^P\right] \quad \text{MP unknowns}$$

$$\boldsymbol{\Theta} = diag\left[\sigma_{\eta_1}^2\ \sigma_{\eta_2}^2 \cdots \sigma_{\eta_M}^2\right] \quad \text{M unknowns}$$

Thus, the likelihood function of $(\mathbf{w}, \boldsymbol{\theta}, \text{and } \boldsymbol{\Theta})$ is given by

$$f(\mathbf{w}, \boldsymbol{\theta}, \Theta) = \frac{1}{\left(\pi^M \prod_{m=1}^{M} \sigma_{\eta_m}^2\right)^{N/2}} \exp\left(-\sum_{k=0}^{N/2-1}\left(\mathbf{N}(k)^H \Theta^{-1}\mathbf{N}(k)\right)\right) \tag{5.32}$$

Since the residual is a close approximation of the noise, $\boldsymbol{\Theta}$ can be estimated from the residual,

$$\begin{aligned}\hat{\mathbf{N}}(k) &= \boldsymbol{\Sigma}(k) - \mathbf{D}(k)\left(\mathbf{D}(k)\right)^{\dagger} \boldsymbol{\Sigma}(k) \\ &= \left(\mathbf{I} - \mathbf{D}(k)\left(\mathbf{D}(k)\right)^{\dagger}\right)\boldsymbol{\Sigma}(k)\end{aligned} \tag{5.33}$$

which is close to the actual noise. Then, the noise variance may be estimated as

$$\hat{\sigma}_{\eta_m}^2 = \frac{2}{N}\sum_{k=0}^{N/2-1}\left|\left\{\hat{N}(k)\right\}_m\right|^2 \tag{5.34}$$

The covariance matrix of estimated noise is

$$\hat{\Theta} = diag[\hat{\sigma}_{\eta_m}^2]$$

$$\begin{aligned}\log f(\mathbf{w},\boldsymbol{\theta}) &= -\frac{N}{2}\sum_{m=1}^{M}\log\hat{\sigma}_{\eta_m}^2 - \sum_{k=0}^{N/2-1}\sum_{m=1}^{M}\frac{\left|[\hat{N}(k)]_m\right|^2}{\hat{\sigma}_{\eta_m}^2} \\ &= -\frac{N}{2}\sum_{m=1}^{M}\log\left\{\frac{2}{N}\sum_{k=0}^{N/2-1}\left|[\hat{N}(k)]_m\right|^2\right\} - M\frac{N}{2} \\ &= \frac{N}{2}M\log\frac{N}{2} - \sum_{m}^{M}\log\left|\hat{N}_m\right|^2 - M\frac{N}{2} \\ &= \frac{N}{2}\left[M\left(\log\frac{N}{2}-1\right)\right] - \sum_{m}^{M}\log\hat{N}_m\end{aligned} \tag{5.35}$$

where $\hat{N}_m = 2/N\sum_{k=0}^{N/2-1}\left|[\hat{N}(k)]_m\right|^2$. To obtain the previous result, we have used Equation 5.34 in Equation 5.35. Essentially, we now need to minimize $\sum_m^M \log\hat{N}_m$ with respect to $\mathbf{w}$ and $\boldsymbol{\theta}$. There are 2MP unknowns in total [9]. Recall that we have defined $\hat{N}(k)$ in Equation 5.33. The unknown parameters are embedded in $\hat{N}(k)$.

For a single transmitter, we need to minimize

$$\underset{\mathbf{w}^1,\boldsymbol{\theta}^1}{\arg\min}\ \sum_{m=1}^{M}\sum_{k=0}^{N/2-1}\left\|B(k,\mathbf{w}^1,\boldsymbol{\theta}^1)\right\|^2 \tag{5.36}$$

where

$$\mathbf{B}(k,\mathbf{w}^1,\boldsymbol{\theta}^1) = \left[(\mathbf{I}-\mathbf{d}^1(k,\mathbf{w}^1,\boldsymbol{\theta}^1)\mathbf{d}^1(k,\mathbf{w}^1,\boldsymbol{\theta}^1)^H\right]\boldsymbol{\Sigma}^1(k)$$

and

$$\mathbf{d}^1(k,\mathbf{w}^1,\boldsymbol{\theta}^1) = \frac{1}{\sqrt{M}}\begin{bmatrix} w_1^1\exp\left(-j\frac{2\pi}{N}k\tau_1^1\right) & w_2^1\exp\left(-j\frac{2\pi}{N}k\tau_2^1\right)\cdots \\ w_M^1\exp\left(-j\frac{2\pi}{N}k\tau_M^1\right) & \end{bmatrix}^T \tag{5.37}$$

For a single transmitter, Equation 5.36 may be simplified as follows:

$$\begin{aligned}\left\|\mathbf{B}(k,\mathbf{w}^1,\boldsymbol{\theta}^1)\right\|^2 &= \left\|\left[(\boldsymbol{\Sigma}^1(k)-\mathbf{d}^1(k,\mathbf{w}^1,\boldsymbol{\theta}^1)\mathbf{d}^1(k,\mathbf{w}^1,\boldsymbol{\theta}^1)^H\boldsymbol{\Sigma}^1(k)\right]\right\|^2 \\ &= \left\|\boldsymbol{\Sigma}^1(k)\right\|^2 - \boldsymbol{\Sigma}^1(k)^H\mathbf{d}^1(k,\mathbf{w}^1,\boldsymbol{\theta}^1)\mathbf{d}^1(k,\mathbf{w}^1,\boldsymbol{\theta}^1)^H\boldsymbol{\Sigma}^1(k) \\ &\quad - \boldsymbol{\Sigma}^{1^H}(k)\mathbf{d}^1(k,\mathbf{w}^1,\boldsymbol{\theta}^1)\mathbf{d}^1(k,\mathbf{w}^1,\boldsymbol{\theta}^1)^H\boldsymbol{\Sigma}^1(k) \\ &\quad + \boldsymbol{\Sigma}^{1^H}(k)^H\mathbf{d}^1(k,\mathbf{w}^1,\boldsymbol{\theta}^1)\mathbf{d}^1(k,\mathbf{w}^1,\boldsymbol{\theta}^1)^H\boldsymbol{\Sigma}^1(k) \\ &= \left\|\boldsymbol{\Sigma}^1(k)\right\|^2 - \left\|\mathbf{d}^1(k,\mathbf{w}^1,\boldsymbol{\theta}^1)^H\boldsymbol{\Sigma}^1(k)\right\|^2\end{aligned} \tag{5.38}$$

The final result consists of two square terms. The first one is independent of the unknown terms, hence, it does not take part in minimization. The second term is of interest. As it has a negative sign, it is enough to maximize the negative term so that Equation 5.36 is minimized.

$$\underset{\mathbf{w}^1,\boldsymbol{\theta}^1}{\arg\max} \sum_{m=1}^{M}\sum_{k=0}^{N/2-1}\left\|\mathbf{d}^1(k,\mathbf{w}^1,\boldsymbol{\theta}^1)^H\boldsymbol{\Sigma}^1(k)\right\|^2 \tag{5.39}$$

Equation 5.39 will be useful later in an expectation-minimization (EM) algorithm, where we are able to express the total log likelihood function as the sum of log likelihood functions of individual transmitters. Note that EM stands for expectation-minimization when the log likelihood is positive, otherwise EM stands for expectation-maximization when log likelihood is negative.

To make the optimization procedure simple, we assume that $w_1^1 = w_2^1 \cdots w_M^1 = 1/M$; then we are left with only two location parameters, x_0 and y_0. Equation 5.37 reduces to

$$\mathbf{d}^1(k) = \frac{1}{\sqrt{M}}\begin{bmatrix}\exp\left(-j\frac{2\pi}{N\Delta t}k\tau_1^1\right) & \exp\left(-j\frac{2\pi}{N\Delta t}k\tau_2^1\right)\cdots \\ \exp\left(-j\frac{2\pi}{N\Delta t}k\tau_M^1\right) & \end{bmatrix}^T \tag{5.40}$$

where $\tau_m^1 = 1/c\sqrt{(x_m - x_0)\wedge 2 + (y_m - y_0)\wedge 2}$.

A simple-minded approach to find the minimum is to compute Equation 5.36 over a grid of points around the expected location of the transmitter. (A more effective algorithm called "Approximately concentrated ML algorithm, AC–ML" is described in [9].) Then, look for the position of the minimum. Given the estimates of the transmitter location, $(\hat{x}_0, \hat{y}_0)$, the angle of arrival (measured with respect to the y-axis) is easily obtained as

$$\boldsymbol{\theta}^1 = \tan^{-1}\left(\frac{\hat{x}_0}{\hat{y}_0}\right) \tag{5.41}$$

The mathematical analysis given previously is based on the work of [8,10], with some simplification of analysis and use of consistent notations. In the example that follows, we give results from our program largely with a view of verification, but without any error analysis, which has been extensively covered in the references mentioned earlier.

EXAMPLE 5.2

An array of ten randomly distributed sensors spread over ±50 m is used to locate a transmitter located at (900, 900 m). The transmitter transmits a hopping frequency signal (four randomly selected frequencies in the range 0–5 MHz) with a carrier frequency of 100 MHz. Each receiver recovers the baseband signal (an example of baseband signal is shown in Figure 5.6a) and samples at a rate of 10 MHz. Discrete samples (256) are then transmitted to the anchor node located at (0,0), where all further processing is carried out. The output of each sensor is contaminated with independent random Gaussian noise (zero mean) and variance equal to (signal power/SNR). For simplicity of computation, a single transmitter is assumed. We have computed (5.36) for a series of x-coordinates but keep the y-coordinate fixed (at 900 m) for two different values of SNR (infinity and 10 dB). The results (negative of computed values) are shown in Figures 5.6b,c. The peak representing log likelihood function is sharp, particularly for a high SNR, at the true location of the transmitter.

The presence of noise basically affects the width of the peak. For SNR = 10 (on linear scale), the log likelihood (negative) is shown in Figure 5.6c. The x-coordinate scan for the different number of sensors is shown in Figure 5.7. The peak height increases almost linearly with the number of sensors, as seen in Figure 5.8. The peak location remains unaltered. The transmitter is now located at (500, 500 m).

EXAMPLE 5.3

Consider a 64-sensor randomly distributed array with two transmitters in the far field (see Figure 5.9). This assumption allows us to simplify the steering vector of Equation 5.40 as shown subsequently. The travel time from a transmitter, say the first transmitter to the mth sensor, can be approximated for $(x_m, y_m) \ll r_0$ as

$$\tau_m^1 \approx \frac{r_0}{c} - \frac{1}{c}(x_m \sin(\theta_0) + y_m \cos(\theta_0))$$

The steering vector under this approximation reduces to

$$\mathbf{d}^1(k) = \frac{\exp\left(-j\frac{r_0}{c}\frac{2\pi}{N\Delta t}k\right)}{\sqrt{M}} \begin{bmatrix} \exp\left(j\frac{2\pi}{N\Delta t}\frac{k}{c}(x_1 \sin(\theta_0) + y_1 \cos(\theta_0))\right) \\ \exp\left(j\frac{2\pi}{N\Delta t}\frac{k}{c}(x_2 \sin(\theta_0) + y_2 \cos(\theta_0)\right) \\ \vdots \\ \exp\left(j\frac{2\pi}{N\Delta t}\frac{k}{c}(x_M \sin(\theta_0) + y_M \cos(\theta_0)\right) \end{bmatrix}$$

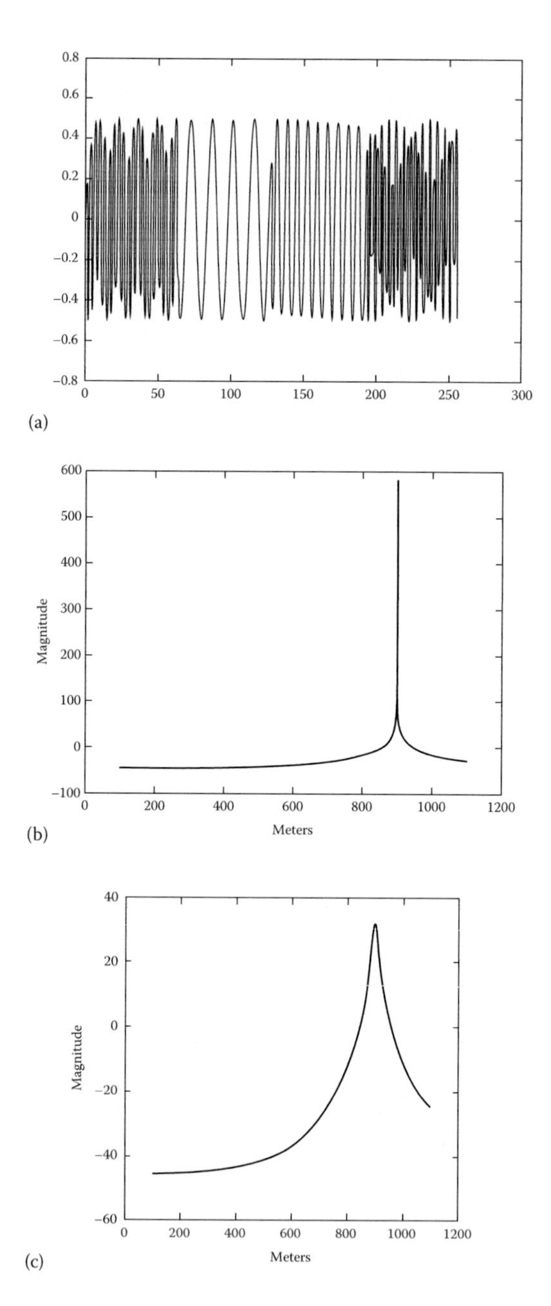

FIGURE 5.6 (a) Hopping frequency signal consisting of four randomly picked frequencies in the range of 0–5 MHz. This forms base-band riding on a carrier signal of 100 MHz. The x-axis is in units of sampling intervals (10^{-7} s). (b) The magnitude is plotted as a function of the x-coordinate starting from 100 m to 1100 m, but for a fixed y-coordinate (=900). The SNR is infinity (i.e., no noise). (c) Same as in (b) but for SNR = 10 dB. With a decreasing SNR, e.g., 0 dB SNR, the peak broadens, but the maximum is located at the correct location.

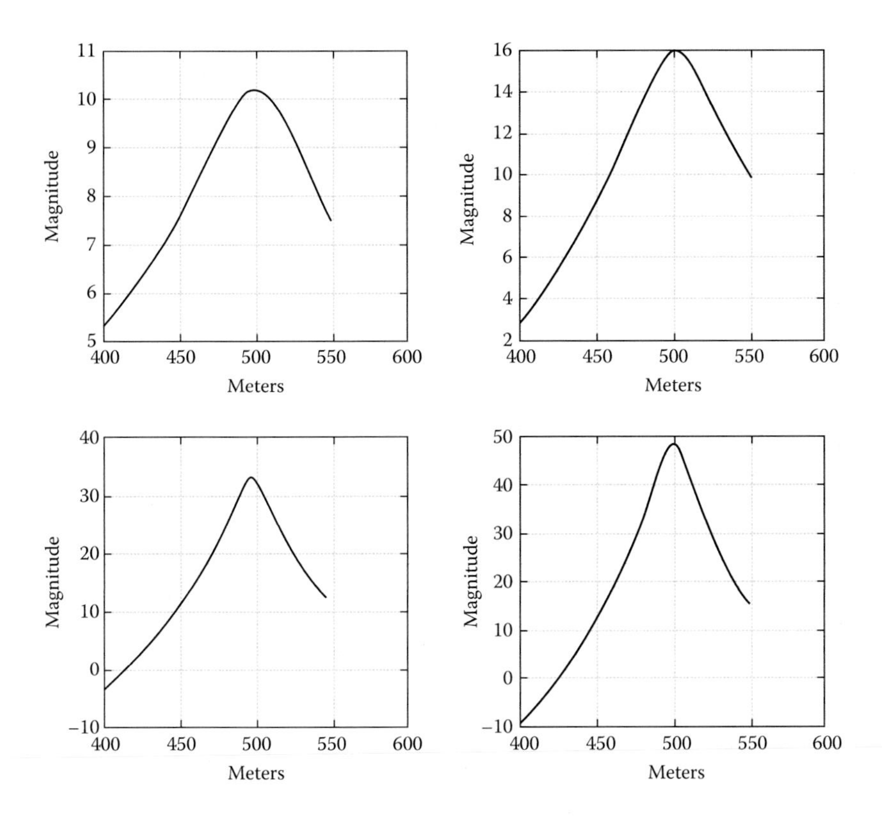

FIGURE 5.7 Scan along x-coordinate for different numbers of sensors, namely, 3, 5, 10, and 15. Other parameters are as in Figure 5.6. Since the sensors are placed randomly each time, the scan changes. But peak height and its location remain unaltered.

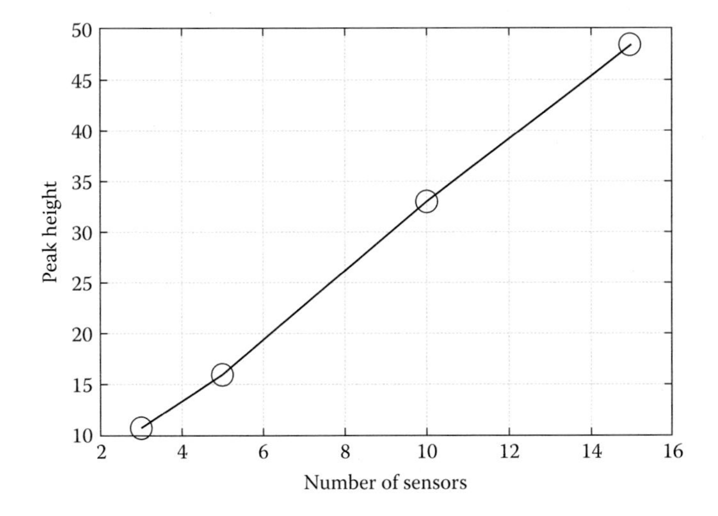

FIGURE 5.8 Peak height increases almost linearly with number of sensors.

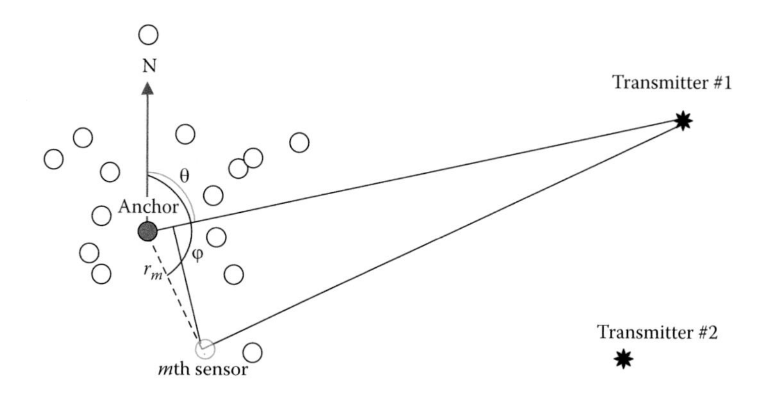

FIGURE 5.9 A 16-element (planar) DSA. The anchor node is shown in gray. All angles are measured with respect to the North.

Notice that the range parameter appears only as a constant phase factor. First, we let the DoAs be at 30.0° and 90.0°, and the result is as shown in Figure 5.10. In an effort to further increase the resolution, we have reduced the DoAs to 30.0° and 65.0°, but now the sources could not be resolved at all without increasing the array size. The transmitters could be resolved only when the diameter of the enclosing circle was increased to 300 m with the same number of sensors (Figure 5.11).

5.4 EXPECTATION-MAXIMIZATION ALGORITHM

In a multi-transmitter case and in non-uniform sensor weighting, the maximization problem of Equation 5.35 becomes a non-linear optimization problem requiring large computing power beyond the scope of any DSA. The EM algorithm (here EM stands for expectation-maximization, as the log likelihood given by Equation 5.45a is negative), on the other hand, is an iterative procedure for the estimation of the maximum likelihood (ML) parameters of just one transmitter at a time. The observed signal is modeled as a linear sum of signals from all transmitters, which cannot be observed individually but constitute the complete information. What is observed as being a sum of the signals from all the transmitters is only partial information. We need to estimate the unknown parameters, given the partial information and signal model. In a frequency domain, we express the model as

$$\Sigma(k) = \sum_{p=1}^{P} \mathbf{X}^{p}(k) \tag{5.42a}$$

where each component is a signal received from the pth transmitter plus noise,

$$\mathbf{X}^{p}(k) = \mathbf{d}^{p}(k) S^{p}(k) + \mathbf{N}^{p}(k)$$

$$\Sigma(k) = \sum_{p=1}^{P} \mathbf{d}^{p}(k) S^{p}(k) + \sum_{p=1}^{P} \mathbf{N}^{p}(k) \tag{5.42b}$$

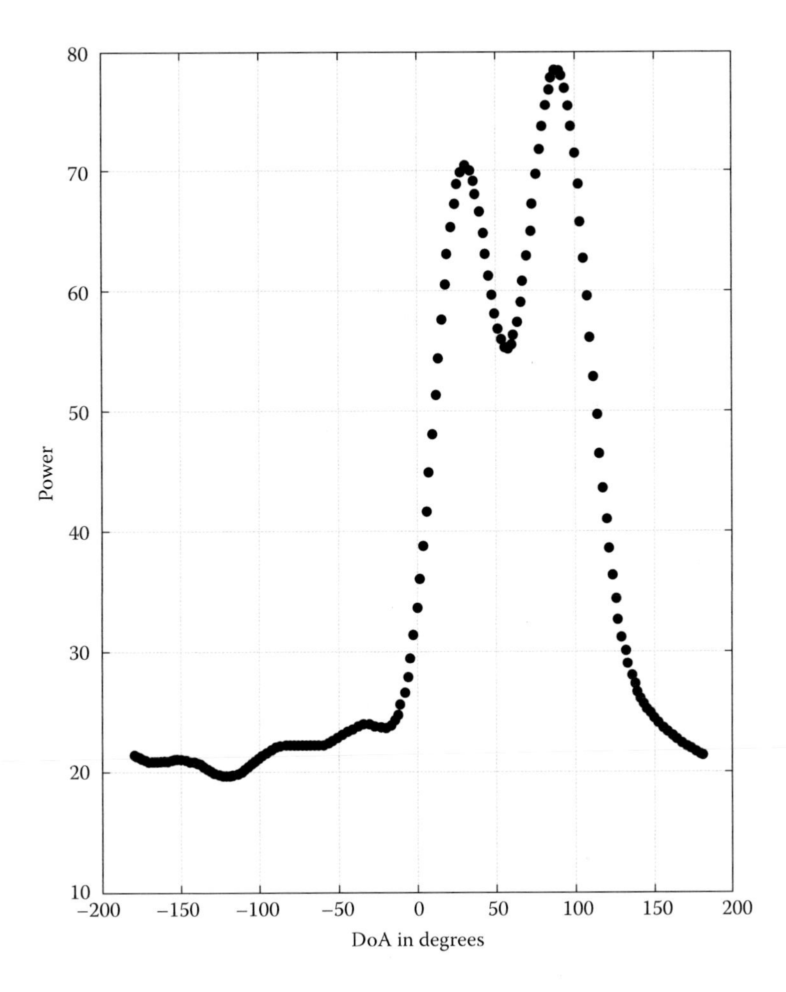

FIGURE 5.10 A 64-element randomly distributed sensor array (DSA) lying inside a circle of diameter 180 m. There are two transmitters in the far field range (= 2000 m). The DoAs of the transmitters are 30.0° and 90.0°. SNR = 10 on a linear scale (10 dB on log scale). The total length of each signal is 8000 points corresponding to 100 μs. The transmitters are well resolved with correct DoA estimates.

where $\mathbf{d}^p(k)$ is defined in Equation 5.40 and $\mathbf{N}^p(k)$ is a complex zero mean uncorrelated Gaussian noise in the sole presence of the pth transmitter. Note that $\mathbf{X}^p(k)$, $\mathbf{d}^p(k)$, and $\mathbf{N}^p(k)$ are all vectors of size $(\mathrm{M} \times 1)$. Initially, at iteration $\mathrm{i} = 0$, we let the noise variance be uniform across all transmitters. Then, the noise covariance matrix is simply given by

$$\Theta^p = \frac{1}{P}\mathbf{I}$$

These are then recursively updated leading to a non-uniform noise. Also, we select initial values for other parameters, namely, $\mathbf{w}$(weights) and θ (DoA or ToA).

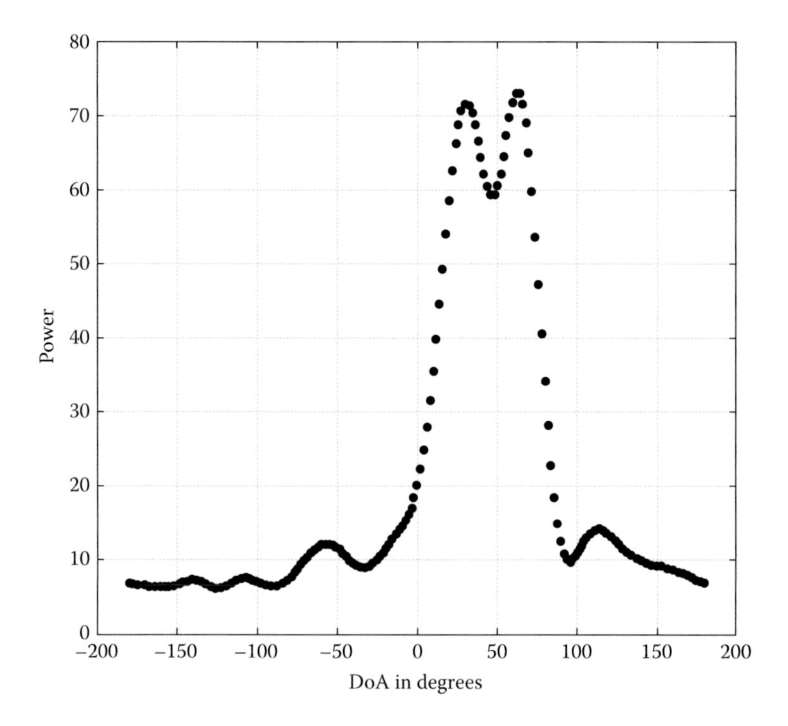

FIGURE 5.11 A 64-element randomly distributed sensor array lying inside a circle of diameter 300 m. There are two transmitters in the far field (range = 2000 m). The DoAs of sources are 30.0° and 65.0°. SNR = 10 on a linear scale (10 dB on log scale). The total length of each signal is 8000 points corresponding to 100 μs. The transmitters are well resolved with correct DoA estimates.

This selection will play a crucial role in arriving at the global maximum. It is recommended that the iterations be initiated from different initial locations in the hope that at least one of the iterations ends up at the global maximum.

5.4.1 Expectation Step (E-Step)

In the ith step, let $\mathbf{X}^p(k)$ be the signal (plus noise) associated with the pth transmitter. Here, we assume that the noise is uniform across the array. Although it is an unrealistic assumption, it makes the algebra simple. Non-uniform noise cases have been worked out recently [12]. The estimate of $\mathbf{X}^p(k)$, given observed data, is

$$\begin{aligned}\hat{\mathbf{X}}^p(k) &= E\left\{\mathbf{X}^p(k) \mid \mathbf{X}(\mathrm{k})\right\}\\ &= \mathbf{d}^p(k)S^p(k) + \frac{1}{P}\left(\Sigma(k) - \mathbf{D}(k)\mathbf{S}(k)\right)\end{aligned} \tag{5.43}$$

In Equation 5.43 we use, in place of $\boldsymbol{S}(k)$, its least-squares estimate from Equation 5.31, we then obtain

$$\hat{\mathbf{X}}^p(k) = E\left\{\hat{\mathbf{X}}^p(k) \mid \Sigma(\mathrm{k})\right\}$$
$$= \mathbf{d}^p(k)S^p(k) + \frac{1}{P}\left(\Sigma(k) - \mathbf{D}(\mathbf{D})^{\dagger}\Sigma(k)\right) \tag{5.44}$$

Note that this form requires the parameters of all transmitters. We are unable to treat each transmitter independently.

The log likelihood function of complete data can be written as

$$\log f_X(\mathbf{w}, \boldsymbol{\theta}, \mathbf{S}) = -\sum_{k=0}^{N/2-1}\sum_{p=1}^{P}\left\|\mathbf{X}^P(k) - \mathbf{d}^p S^p\right\|^2$$
$$= -\sum_{p=1}^{P}\left\{\sum_{k=0}^{N/2-1}\left\|\mathbf{X}^P(k) - \mathbf{d}^p S^p\right\|^2\right\} \tag{5.45a}$$

where the quantity inside the curly brackets in Equation 5.45a is what we need to optimize. It represents the log likelihood function of the pth transmitter (see Equation 5.36). Thus, the log likelihood function of observed (incomplete) data is a sum of individual log likelihood functions.

$$\log f_{\mathrm{X}}(\mathbf{w}, \theta, \mathbf{S}) = \sum_{P-1}^{P} \log f_{\mathbf{X}^p}(w^P, \theta^P, S^P) \tag{5.45b}$$

where

$$\log f_{\mathbf{X}^p}(w^P, \theta^P, \mathbf{S}^P) = -\sum_{k=0}^{N/2-1}\left\|\mathbf{X}^P(k) - \mathbf{d}^p(k, \theta^P)S^p(k)\right\|^2 \tag{5.45c}$$

To optimize the log likelihood function of the complete data it is enough that we optimize individual log likelihood functions over a $N/2+2$ dimensional space, $\gamma^P = \left[w^P, \theta^P, S^p(k), k = 0, 1, \cdots N/2-1\right]$.

Since $\mathbf{X}^P(k)$ is not available, the expectation of $\mathbf{X}^P(k)$ has to be performed using the observations and the current estimates of the parameters. Compute the conditional expectation given the observations, $\mathbf{X}$

$$Q(\gamma^p, \gamma^{p\,[i]}) = E\left\{\log f_{\mathrm{X}}(\gamma^p; \hat{\mathbf{X}}^p(k, \gamma^{p\,[i]})\right\}$$
$$p = 1, \cdots P \tag{5.46a}$$

where

$$\hat{\mathbf{X}}^p(k, \gamma^{p[i]}) = E\left\{\mathbf{X}^P(k) \mid \Sigma(k), \gamma^{p[i]}\right\}$$
$$= \mathbf{d}^p(k)S^p(k) + \frac{1}{P}\left(\Sigma(k) - \mathbf{D}(k)\mathbf{S}(k)\right) \tag{5.46b}$$

It is noted that $\mathbf{d}^p(k)$, $S^P(k)$ (hence, $\mathbf{D}(k)$ and $\mathbf{S}(k)$) are computed using parameters $\gamma^{P[i]}$ from previous iterations. The Q function is the output of the expectation step, first introduced in the seminal paper on the EM algorithm by Dempster et al. [12] in 1977.

5.4.2 Maximization Step (M-Step)

For a single transmitter, the log likelihood function is given by Equation 5.36. It represents noise associated with the transmitter. Compute the average noise power, that is, the average over all the frequency bins

$$\text{Noise power} = \frac{2}{N}\sum_{k=0}^{N/2-1}|g(k)|^2$$

where

$$g = \mathbf{d}(k,\gamma)^H\left(\Sigma(k) - \mathbf{D}(k)\mathbf{S}(k)\right)$$

Maximize the sum with respect to $\mathbf{w}, \theta, S(k)$ for the transmitter

$$\arg\max_{\gamma} \frac{2}{N}\sum_{k=0}^{N/2-1}\left|\mathbf{d}(k,\gamma)^H\left(\Sigma(\mathrm{k})\text{-}\mathbf{D}\left(k,\gamma^{[i]}\right)\mathbf{S}(k)\right)\right|^2 \tag{5.47}$$

$$\Rightarrow \gamma^{[i+1]}$$

For simplicity, we shall assume that $w^p = 1\ \forall p$. We are left with two parameters γ, $S(k)$, whose maximization leads to the following two steps [10]:

$$\gamma^{[i+1]} = \arg\max_{\gamma}\frac{2}{N}\sum_{k=0}^{N/2-1}\left|\mathbf{d}(k,\gamma)^H\left(\Sigma(k) - \mathbf{D}\left(k,\gamma^{[i]}\right)\mathbf{S}(k)\right)\right|^2 \tag{5.48}$$

and

$$\hat{S}(k)^{[i+1]} = \frac{\mathbf{d}(k,\gamma^{[i+1]})^H(\Sigma(k) - \mathbf{D}(k,\gamma^{[i]})\hat{S}(k)^{[i]})}{\left\|\mathbf{d}(k,\gamma^{[i+1]})\right\|^2} \tag{5.49}$$

$$\forall k$$

Instead of DoA, we may choose to minimize with respect to ToA or transmitter location coordinates. In fact, the ToAs are related to DoAs, albeit non-linearly (Equation 5.26). It is also interesting to note that all ToAs may be expressed in terms of the transmitter and sensor locations, of which sensor locations are known. Let (x^p, y^p) be the coordinates of the pth transmitter. The ToA from the transmitter to the mth sensor is given by

$$\tau_m^p = \frac{r_m^p}{c} = \frac{\sqrt{(x_m - x^p)^2 + (y_m - y^p)^2}}{c}, \quad m = 0,1,\cdots M-1 \tag{5.50}$$

At the maximum point, we compute $g_m^p(k)$

$$g_m^p(k) = \left\{\mathbf{d}^p(k,\gamma^p)^H\right\}_m \left\{\hat{\mathbf{X}}^p(k,\gamma^{p\,[i]})\right\}_m$$

and its covariance function is given by Equation 5.35, where we have $\mathrm{N}_m^p(k)$ in place of $g_m^p(k)$. This is the i+1st estimate of the covariance matrix.

$$[\hat{\Theta}]^{[i+1]} = diag\left[\frac{2}{NM}\sum_{p=}^{P}\sum_{k=0}^{N/2-1}\left|g_m^p(k)\right|^2\right] \tag{5.51}$$

where

$$g^p(k,\theta) = \mathbf{d}^p(k,\theta)^H\hat{\mathbf{X}}^p(k,\theta^{p\,[i]})$$

$$= \mathbf{d}^p(k,\theta)^H\mathbf{d}^p(k,\theta^{p[i]})S^{p[i]}(k)$$

$$+\frac{\mathbf{d}^p(k,\theta)^H}{P}(\Sigma(k) - \mathbf{D}(k)\mathbf{S}(k)) \tag{5.52}$$

This completes the i+1st step. The iteration, starting from $i = 1$, is continued until the convergence stabilizes. Corresponding estimates of DoAs or ToAs are the desired parameters. We have also noted that the minimization of Equation 5.51 can also be carried out with respect to (x^p, y^p), straight away avoiding the intermediate step of minimization with respect to ToAs along with Equation 5.50.

EXAMPLE 5.4

As an illustration of an EM algorithm, we consider a DSA consisting of 64 randomly distributed sensors. The sensors are confined within a square the size of $100 \times 100\ \mathrm{m}^2$, and a single transmitter is on the x-axis at a distance of 95 m. The transmitter radiates a known narrow band frequency hopping signal. The center frequency and bandwidth are 100 and 25 MHz, respectively. Each sensor retransmits the received signal back to a centrally located anchor having sufficient computing power. For processing, 1000 time samples were used.

- Number of sensors: 64
- Number of time samples: 1000
- SNR (linear scale): 10

- Single transmitter is located
- On the x-axis at a distance: 95 m
- All sensors lie inside 100 x 100m^2
- Rectangular space in the x–y plane

Results of the iterative estimation of the range along the x-axis (keeping the y-axis range fixed, equal to zero) are shown in Table 5.2.

5.4.3 Estimation of Transmitted Signal

Apart from the estimation of the DoA of a transmitter, we would also be interested in what is being transmitted, for example, a bit stream representing a communication signal source, a strategic command, and so on. We show how this can be achieved in the presence of another interfering transmitter. Equations 5.42a and 5.42b may be combined into a single equation

$$\Sigma(k) = \sum_{p=1}^{2} \mathbf{X}^{p}(k)$$
$$= \mathbf{d}^{1}(k)S^{1}(k) + \mathbf{d}^{2}(k)S^{2}(k) + \mathbf{N}(k)$$

where the steering vectors $\mathbf{d}^1$and $\mathbf{d}^2$ are completely defined by DoA. When the steering vectors are orthogonal, the transmitted signal is given by Equation 5.49. But the steering vectors become orthogonal only for a uniformly spaced linear array. We have shown in Section 4.3.3, that it is possible to design an annihilating matrix for any given vector. Given the DoA, we can define the steering vector and corresponding annihilating matrix. Let $\mathbf{P}^1$ and $\mathbf{P}^2$ be the annihilating matrices corresponding to steering vectors $\mathbf{d}^1$and $\mathbf{d}^2$, respectively. To annihilate the first source, pre-multiply the data vector $\Sigma(k)$ with $\mathbf{P}^1$. We obtain

$$\mathbf{P}^{1}\Sigma(k) = \mathbf{P}^{1}\mathbf{d}^{2}(k)S^{2}(k) + \mathbf{P}^{1}\mathbf{N}(k) \tag{5.53}$$

TABLE 5.2
Estimated Range Stabilizes on Second Iteration

Iteration	Range Estimate (m)
1	95.1
2	95.070
3	95.070
4	95.071

Note: Error is less than 0.06%

The least-squares estimate of $S^2(k)$ is given by

$$S^2(k) = \frac{(\mathbf{P}^1\mathbf{d}^2(k))^H \mathbf{P}^1\Sigma(k)}{(\mathbf{P}^1\mathbf{d}^2(k))^H \mathbf{P}^1\mathbf{d}^2(k)} \tag{5.54a}$$

Similarly, $S^1(k)$ may be obtained:

$$S^1(k) = \frac{(\mathbf{P}^2\mathbf{d}^1(k))^H \mathbf{P}^2\Sigma(k)}{(\mathbf{P}^2\mathbf{d}^1(k))^H \mathbf{P}^2\mathbf{d}^1(k)} \tag{5.54b}$$

Inversing the fast Fourier transform (FFT) of $S^1(k)$ and $S^2(k)$ will then yield an estimate of the transmitted signals. The mean square error is very small, of the order of 10^{-3} for parameters in Figure 5.11.

5.5 LOCALIZATION IN A MULTIPATH ENVIRONMENT

There are practical situations where a transmitter is in close proximity to reflecting objects such as the walls of a room. Both the transmitter and DSA are placed inside the room. This situation arises, for example, when a speaker is addressing a conference in a room and it is required to track the position of the speaker. Another case is that of a shallow sea where both the top surface and bottom act as good reflectors. Consider a simple model where a transmitter and DSA lie between two parallel reflecting surfaces. This model can represent a speaker between two walls of a long corridor or an acoustic transmitter in a shallow sea bounded by an air–water interface and a smooth, hard bottom.

5.5.1 Time Domain

The transmitter signal will reach a sensor via M different paths, which differ from each other in delay and attenuation. The sensor output may now be modeled as follows:

$$s(t) = \sum_{m=1}^{M} a_m s_0(t - \tau_m) \tag{5.55}$$

where a_m and τ_m are the attenuation and delay of the mth path, respectively, and $s_0(t)$ is the transmitted signal, which is assumed to be known. We sample the transmitted signal starting with a delay τ, and the sampling interval is Δt. Similarly, we sample the observed signal but without any delay. We call these the pth snapshots:

A snapshot of a transmitted signal after delay:

$$\mathbf{s}(\tau) = \begin{bmatrix} s_0(t-\tau), s_0(t-\tau+\Delta t), s_0(t-\tau+2\Delta t) \\ \cdots s_0(t-\tau+(N-1)\Delta t) \end{bmatrix} \tag{5.56}$$

and a snapshot of the observed signal:

$$\mathbf{y} = \begin{bmatrix} \sum_{m=1}^{M} a_m s_0(t-\tau_m), & \sum_{m=1}^{M} a_m s_0(t-\tau_m+\Delta t), \\ \cdots & \sum_{m=1}^{M} a_m s_0(t-\tau_m+(N-1)\Delta t) \end{bmatrix} \tag{5.57}$$

Let us now form an inner product of the previous two snapshot vectors defined in Equations 5.56 and 5.57, as a function of delay, τ.

$$\varepsilon(\tau) = \frac{1}{N}\mathbf{y}^T\mathbf{s}(\tau) =$$

$$\frac{1}{N}\begin{bmatrix} \sum_{m=1}^{M} a_m s_0(t-\tau_m)s_0(t-\tau) + \sum_{m=1}^{M} a_m s_0(t-\tau_m+\Delta t)s_0(t-\tau+\Delta t) \\ \cdots + \sum_{m=1}^{M} a_m s_0(t-\tau_m+(N-1)\Delta t)s_0(t-\tau+(N-1)\Delta t) \end{bmatrix} \tag{5.58}$$

where (T) stands for transpose. When the transmitted signal is uncorrelated, with a correlation distance less than $\min\{\tau_m, m=1,\ldots M\}$, Equation 5.58 reduces to a simple result

$$\left.\begin{aligned} &\varepsilon(\tau)\big|_{\tau=\tau_m} = a_m \frac{1}{N}\sum_{n=0}^{N-1} s_p^2(n\Delta t) \\ &\text{and} \\ &\varepsilon(\tau)\big|_{\tau\neq\tau_m} = 0 \end{aligned}\right\} \tag{5.59}$$

Thus, the peaks of the product, computed as a function of delay, correspond to delays of multipaths. The height of a peak is proportional to the amplitude.

$$\varepsilon(\tau)\big|_{\tau=\tau_m} = a_m\gamma$$

where

$$\gamma = \frac{1}{N}\sum_{n=0}^{N-1} s_p^2(n\Delta t)$$

For a large N, γ shall be practically independent of τ and τ_m, but it will depend upon the waveform. For example, for a frequency hopping signal, it lies between

0.37925 to 0.37753 with an initial delay of $\tau - \tau_m = 10$ sample units and data length = 100, 500, and 900. Note that the previous statement does not however apply to a correlated waveform such as a narrow band signal.

We can form more than one snapshot by simply changing the starting time. Assuming that the transmitter is stationary and the environment is stable over an interval of time, we can form P independent snapshots. Then, average the product defined in Equation 5.58 over all snapshots. This step will reduce the effect of noise; particularly, all spurious peaks, if any, will be reduced in magnitude. The previous approach uses just one sensor. All other sensors in a DSA may carry out the same operations in parallel and communicate the delay estimates to the anchor node, thus requiring a low communication overhead. In this set-up, however, each sensor will require some computing power and basic computational capabilities, such as, addition, multiplication, and memory. In [13], a subspace approach is described, but it requires the eigenvalue decomposition of a signal matrix. While the subspace approach seems to provide better results, the computational overheads are beyond the number-crunching capacity of most DSAs.

Delay estimation alone will not result in transmitter localization. We need to have extra knowledge of the propagation properties of the channel. This becomes a critical issue in underwater localization because propagation speed varies spatially. One needs to resort to ray tracing (with ray displacement corrections [14,15]) to compute different paths and their delays. Fortunately, for speaker localization in a room or localization of a mobile phone in an urban environment, where wave speed is constant, the principles of geometric ray optics are readily applicable. It is then easy to trace all possible paths having significant power and to compute their delays. In fact, this approach is used to compute paths from every point in the space around the transmitter. Place a test transmitter at point $\mathbf{r} = (x_s, y_s, z_s)$ and trace all possible M significant paths ending in the sensor. In Figure 5.12, we show a transmitter and sensor between two reflecting surfaces (e.g., shallow sea). From knowledge of the propagation environment, we estimate amplitude $a_m(\mathbf{r})$ and delay $\tau_m(\mathbf{r})$ for all M paths. As with the snapshot defined in Equation 5.56, we define a model snapshot as

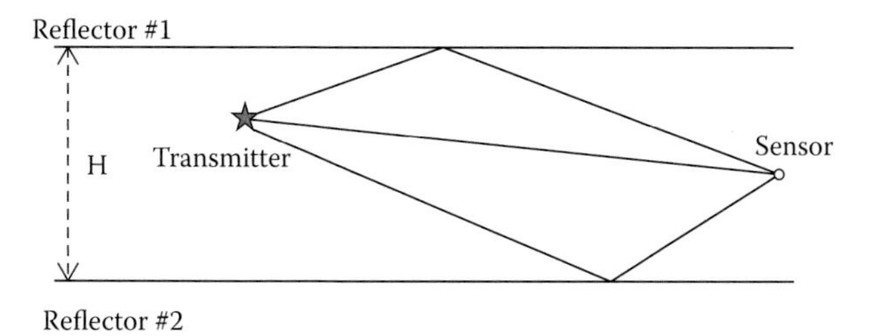

FIGURE 5.12 A transmitter and sensor between two parallel reflecting surfaces. There are three paths: one direct and two reflected paths. All other multiple reflections are not considered, as their strength is likely to be insignificant.

$$\mathbf{s}(\mathbf{r}) = \begin{bmatrix} \sum_{p=1}^{P} a_p(\mathbf{r}) s_0(t-\tau_p(\mathbf{r})), \sum_{p=1}^{P} a_p(\mathbf{r}) s_0(t-\tau_p(\mathbf{r})+\Delta t), \\ \cdots \sum_{p=1}^{P} a_p(\mathbf{r}) s_0(t-\tau_p(\mathbf{r})+(N-1)\Delta t) \end{bmatrix} \tag{5.60}$$

The dot product between two vectors (defined in Equation 5.58) becomes

$$\begin{aligned} \varepsilon(\mathbf{r}) &= \frac{1}{N}\mathbf{y}^T\mathbf{s}^*(\mathbf{r}) \\ &= \frac{1}{N}\begin{bmatrix} \sum_{p=1}^{P} a_p s_0(t-\tau_p) \sum_{p=1}^{P} a_p^*(\mathbf{r}) s_0(t-\tau_p(\mathbf{r})) + \quad \cdots \\ + \sum_{p=1}^{P} a_p s_0(t-\tau_p+(N-1)\Delta t) \sum_{p=1}^{P} a_p^*(\mathbf{r}) s_0(t-\tau_p(\mathbf{r})+(N-1)\Delta t) \end{bmatrix} \end{aligned} \tag{5.61}$$

In a special case where the transmitted signal is uncorrelated (correlation distance tending to zero), Equation 5.61 reduces to

$$\begin{aligned} \varepsilon(\mathbf{r})\big|_{\mathbf{r}=\mathbf{r}_0} &= \sum_{p=1}^{P} |a_p|^2 \sum_{n=1}^{N-1} s_0^2(t-\tau_p+n\Delta t) \\ &= \sum_{p=1}^{P} |a_p|^2 \gamma \\ \varepsilon(\mathbf{r})\big|_{\mathbf{r}\neq\mathbf{r}_0} &= 0 \end{aligned} \tag{5.62}$$

where γ is given by

$$\gamma = \frac{1}{N}\sum_{n=0}^{N-1} s_0^2(t-\tau_p+n\Delta t)$$

It is a measure of the likelihood of the transmitter being present at $\mathbf{r}$. Wherever $\varepsilon(\mathbf{r})$ peaks, it is the actual location of the transmitter. Theoretically, just one sensor is enough for localization, as in [13], when the transmitted signal is truly broadband.

EXAMPLE 5.5

We shall consider a simple illustration of the previous approach. Let there be a transmitter and a single sensor between two parallel reflecting surfaces. The propagation speed is constant, an idealized model of a shallow-water sea channel (see Figure 5.12).

We place the center of coordinates on reflector #1 just above the sensor. The positive z-coordinate points downward and the positive x-axis points left (toward the transmitter). The sensor location is at $(0, z_s)$ and the transmitter location is at (x_e, z_e). The lengths of three paths are as follows:

Direct path: $\sqrt{x_e^2 + (z_e - z_s)^2}$
via reflector #1: $\sqrt{x_e^2 + (z_e + z_s)^2}$
via reflector #2: $\sqrt{x_e^2 + (2H - z_e - z_s)^2}$

The delays suffered by the three paths are

$$\tau_1 = \sqrt{x_e^2 + (z_e - z_s)^2}/c, \quad \tau_2 = \sqrt{x_e^2 + (z_e + z_s)^2}/c, \text{ and } \tau_3 = \sqrt{x_e^2 + (2H - z_e - z_s)^2}/c.$$

Let path loss be a_1, a_2, and a_3, which includes loss due to reflection, medium attenuation, geometrical spreading, and so on. The observed signal vector is

$$\mathbf{s}(t) = \begin{bmatrix} \sum_{p=1}^{3} a_p s_0(t-\tau_p), & \sum_{p=1}^{3} a_p s_0(t-\tau_p+\Delta t), \\ \cdots & \sum_{p=1}^{3} a_p s_0(t-\tau_p+(N-1)\Delta t) \end{bmatrix}$$

and the model vector is given by

$$\mathbf{s}(\mathbf{r}) = \begin{bmatrix} \sum_{p=1}^{3} a_p(\mathbf{r}) s_0(t-\tau_p(\mathbf{r})) & \sum_{p=1}^{3} a_p(\mathbf{r}) s_0(t-\tau_p(\mathbf{r})+\Delta t) \\ \cdots & \sum_{p=1}^{3} a_p(\mathbf{r}) s_0(t-\tau_p(\mathbf{r})+(N-1)\Delta t) \end{bmatrix}$$

The dot product of the observed vector and model vector given by Equation 5.60 reduces, as shown subsequently, where we have used the normalized dot product with respect to the data length

$$\varepsilon(\mathbf{r})\Big|_{\mathbf{r}=\mathbf{r}_0} = \frac{1}{N}\left[\sum_{p=1}^{3} |a_p|^2 \sum_{n=0}^{N-1} s_0^2(t-\tau_p+n\Delta t)\right]$$

$$\varepsilon(\mathbf{r})\Big|_{\mathbf{r}\neq\mathbf{r}_0} = 0$$

The transmitter transmits a frequency hopping signal with 64 random frequencies with random phases lying within a frequency band of ± 0.25 Hz (being normalized, the maximum frequency is 0.5 Hz). If the transmitted signal is known to the receiver, then only we will be able to build a model vector for a given propagation environment. The attenuation coefficients were assumed to be known, which are $a_1 = 1.0$, $a_2 = 0.8$, and $a_3 = 0.6$. The total data length is 4096 points with a sampling interval of 0.5 ms (bandwidth = 2000 Hz centered at 1000 Hz). The data

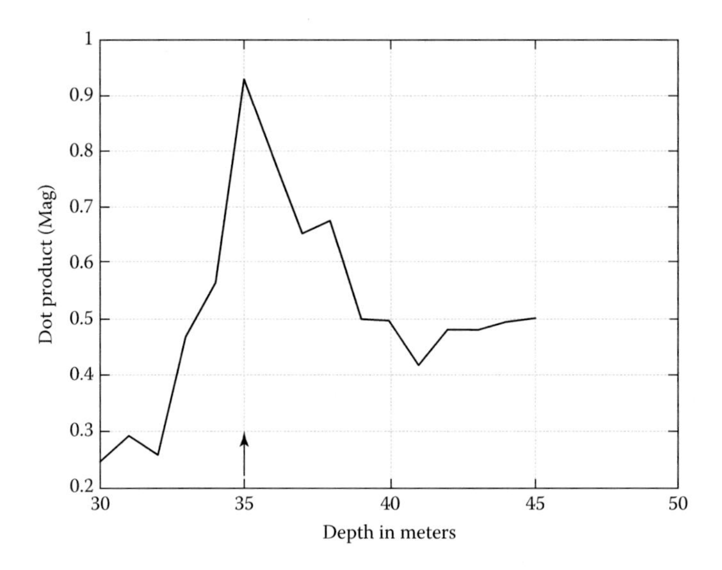

FIGURE 5.13 Dot product as a function of depth for a fixed range (200 m). The arrow indicates the position of the transmitter.

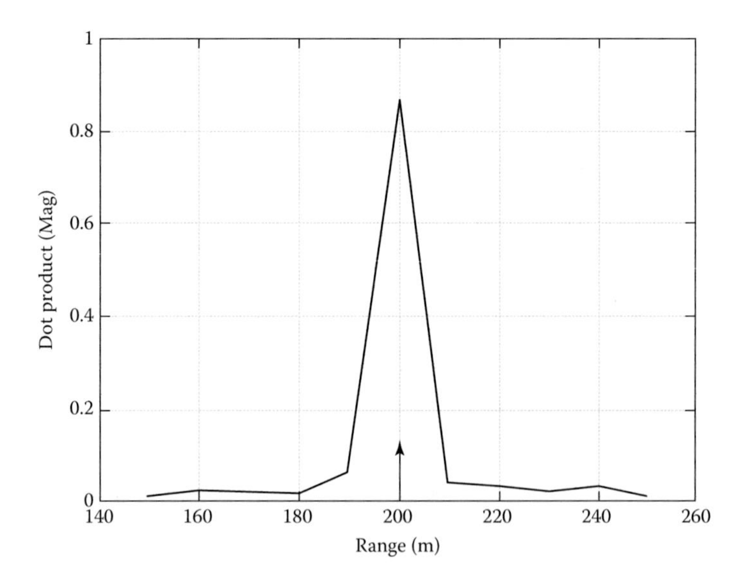

FIGURE 5.14 Dot product as a function of range for a fixed depth (35 m). The arrow indicates the position of the transmitter.

length must be much longer than the tau_{max} so that γ defined in Equation 5.62 is independent of propagation delays. Since the signal is real, the spectrum will be reflected onto the negative frequency axis. For simplicity, we have computed the dot product (magnitude) as a function of the range or depth keeping the other parameter fixed and known. The results are shown in Figures 5.13 and 5.14. The maximum is sharp and located at the correct transmitter position. These results were obtained for snr = 10 db. The range estimation seems to be more precise than the depth estimation, as the depth scan (Figure 5.13) has a broader peak.

5.5.2 Frequency Domain

The critical requirement to enable localization in a confined space with a single sensor is that the transmitted signal must be uncorrelated (i.e., broadband). But with multiple sensors, we show how to overcome this limitation. Consider an M sensor DSA in a confined space (i.e., a room). There are P (known) multipaths between the transmitter and sensor. The transmitter transmits a narrowband (acoustic signal). A sum of P replicas of the transmitted signal, that has been delayed and attenuated, reaches the ith sensor. Following Equation 5.55, we can express the received signal as

$$s(t) = \sum_{p=1}^{P} a_p s_0(t - \tau_p) \tag{5.63}$$

We take the DFT of the output of each sensor. The collection of all outputs may be expressed in a matrix form as

$$\underset{M\times 1}{\begin{bmatrix} S_1(\omega) \\ S_2(\omega) \\ \vdots \\ S_M(\omega) \end{bmatrix}} = \underset{M\times P}{\begin{bmatrix} a_1^1 e^{-j\tau_1^1\omega} & a_2^1 e^{-j\tau_2^1\omega} \cdots & a_P^1 e^{-j\tau_P^1\omega} \\ a_1^2 e^{-j\tau_1^2\omega} & a_2^2 e^{-j\tau_2^2\omega} \cdots & a_P^2 e^{-j\tau_P^2\omega} \\ & & \\ a_1^M e^{-j\tau_1^M\omega} & a_2^M e^{-j\tau_2^M\omega} \cdots & a_P^M e^{-j\tau_P^M\omega} \end{bmatrix}} \underset{P\times 1}{\begin{bmatrix} S_0(\omega) \\ S_0(\omega) \\ \vdots \\ S_0(\omega) \end{bmatrix}} \tag{5.64}$$

$$\mathbf{S} = \mathbf{\Gamma}\mathbf{S}_0$$

where $S_0(\omega)$ is the DFT of $s_0(t)$ and similarly $S_1(\omega)$, $S_2(\omega)$, and so on, are the DFT of sensor outputs. We select frequency ω corresponding to a large signal power. So far, we have not specified the propagation channel. But, we do that now and compute the expected received signals. This is our model sensor output, where the overhead tilde stands for the model parameter:

$$\begin{bmatrix} \tilde{S}_1(\omega) \\ \tilde{S}_2(\omega) \\ \vdots \\ \tilde{S}_M(\omega) \end{bmatrix} = \begin{bmatrix} \tilde{a}_1^1 e^{-j\tilde{\tau}_1^1\omega} & \tilde{a}_2^1 e^{-j\tilde{\tau}_2^1\omega} \cdots & \tilde{a}_P^1 e^{-j\tilde{\tau}_P^1\omega} \\ \tilde{a}_1^2 e^{-j=\tilde{\tau}_1^2\omega} & \tilde{a}_2^2 e^{-j\tilde{\tau}_2^2\omega} \cdots & \tilde{a}_P^2 e^{-j\tilde{\tau}_P^2\omega} \\ & & \\ \tilde{a}_1^M e^{-j\tilde{\tau}_1^M\omega} & \tilde{a}_2^M e^{-j\tilde{\tau}_2^M\omega} \cdots & a\tilde{a}_P^M e^{-j\tilde{\tau}_P^M\omega} \end{bmatrix} \begin{bmatrix} S_0(\omega) \\ S_0(\omega) \\ \vdots \\ S_0(\omega) \end{bmatrix} \tag{5.65}$$

$$\tilde{\mathbf{S}} = \tilde{\mathbf{\Gamma}}\mathbf{S}_0$$

We define a dot product of the observed sensor output vector and model output vector:

$$\tilde{\mathbf{S}}\mathbf{S}^H = \tilde{\mathbf{\Gamma}}\mathbf{S}_0\mathbf{S}_0^H\mathbf{\Gamma}^H$$

$$= \left|\mathbf{S}_0\right|^2 \tilde{\mathbf{\Gamma}}\boldsymbol{\phi}\mathbf{\Gamma}^H \tag{5.66}$$

where ϕ is a $p \times p$ matrix whose all elements are ones. The quantity of interest is $\tilde{\mathbf{\Gamma}}\boldsymbol{\phi}\mathbf{\Gamma}^H$, particularly when the test location lies at the actual transmitter location, that is, $\mathbf{r} = \mathbf{r}_0$. It turns out that at $\mathbf{r} = \mathbf{r}_0$, all diagonal elements of $\tilde{\mathbf{\Gamma}}\boldsymbol{\phi}\mathbf{\Gamma}^H$ are real and equal to 1. This property can be used to determine when the test transmitter has reached the correct location. In an actual computation, we evaluate $\tilde{\mathbf{S}}\mathbf{S}^H$, on account of Equation 5.66, the diagonal terms shall be equal to $\left|\mathbf{S}_0\right|^2$, the spectrum of the transmitted signal. We illustrate the previous algorithm through an example.

EXAMPLE 5.6

We consider an example of a shallow sea, as in Example 5.5, with a modification where we have five sensors suspended from floating sonobuoys, as shown in Figure 5.15. The sensor locations (x_i, y_i, z_i $\text{l} = 1, 2, \ldots, M$) are known. We place the center of the coordinates on reflector #1 just above the anchor node, which is at (0, 0, z_a). All other sensors are in a direct communication link with the anchor node. The location of the transmitter (x_e, y_e, z_e) is also measured with reference to the anchor node. For simplicity, we consider only three paths: the direct path, the first from the top reflector, and the second from the bottom reflector. The path lengths are

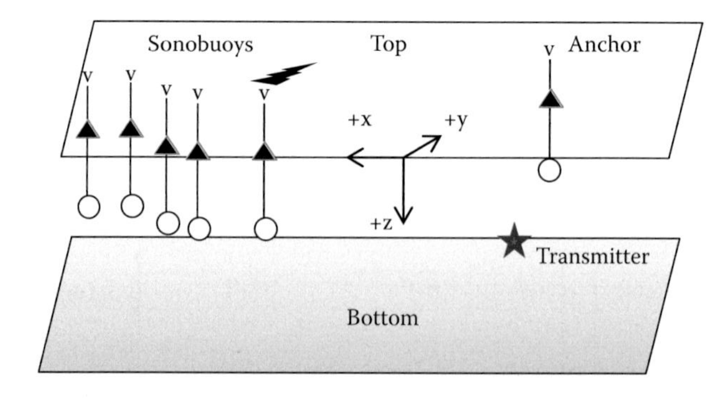

FIGURE 5.15 Sensors and transmitter between two reflectors (e.g., shallow sea). The anchor node receives the transmission from an individual sensor via the transmitter mounted on the sonobuoy.

Direct path: $\sqrt{(x_e - x_i)^2 + (y_e - y_i)^2 + (z_e - z_i)^2}$
via reflector #1: $\sqrt{(x_e - x_i)^2 + (y_e - y_i)^2 + (z_e + z_i)^2}$
via reflector #2: $\sqrt{(x_e - x_i)^2 + (y_e - y_i)^2 + (2H - z_e - z_i)^2}$

The observed signal vector is

$$\mathbf{s}_i(t) = \begin{bmatrix} \sum_{p=1}^{3} a_p^i s_0(t - \tau_p^i) & \sum_{p=1}^{3} a_p^i s_0(t - \tau_p^i + \Delta t) \\ \cdots & \sum_{p=1}^{3} a_p^i s_0(t - \tau_p^i + (N-1)\Delta t) \end{bmatrix} \quad i = 1, 2, \cdots M$$

and its Fourier transform is

$$S_i(\omega) = \left[\sum_{p=1}^{3} a_p^i e^{-j\tau_p^i \omega} S_0(\omega) \right] \quad i = 1, 2, \cdots M$$

We have suppressed variable t as it is the starting time instant from where the next N samples are derived for DFT computation. Collecting all Fourier coefficients from M sensors in a vector form, we get a matrix equation:

$$\underset{M \times 1}{\begin{bmatrix} S_1(\omega) \\ S_2(\omega) \\ \vdots \\ S_M(\omega) \end{bmatrix}} = \underset{M \times 3}{\begin{bmatrix} a_1^1 e^{-j\tau_1^1 \omega} & a_2^1 e^{-j\tau_2^1 \omega} & a_3^1 e^{-j\tau_3^1 \omega} \\ a_1^2 e^{-j\tau_1^2 \omega} & a_2^2 e^{-j\tau_2^2 \omega} & a_3^2 e^{-j\tau_3^2 \omega} \\ & & \\ a_1^M e^{-j\tau_1^M \omega} & a_2^M e^{-j\tau_2^M \omega} & a_3^M e^{-j\tau_3^M \omega} \end{bmatrix}} \underset{3 \times 1}{\begin{bmatrix} S_0(\omega) \\ S_0(\omega) \\ S_0(\omega) \end{bmatrix}}$$

We do the same operations on the model vector and obtain an equivalent expression:

$$\begin{bmatrix} \tilde{S}_1(\omega) \\ \tilde{S}_2(\omega) \\ \vdots \\ \tilde{S}_M(\omega) \end{bmatrix} = \begin{bmatrix} \tilde{a}_1^1 e^{-j\tilde{\tau}_1^1 \omega} & \tilde{a}_2^1 e^{-j\tilde{\tau}_2^1 \omega} & \tilde{a}_3^1 e^{-j\tilde{\tau}_3^1 \omega} \\ \tilde{a}_1^2 e^{-j\tilde{\tau}_1^2 \omega} & \tilde{a}_2^2 e^{-j\tilde{\tau}_2^2 \omega} & \tilde{a}_3^2 e^{-j\tilde{\tau}_3^2 \omega} \\ & & \\ \tilde{a}_1^M e^{-j\tilde{\tau}_1^M \omega} & \tilde{a}_2^M e^{-j\tilde{\tau}_2^M \omega} & \tilde{a}_3^M e^{-j\tilde{\tau}_3^M \omega} \end{bmatrix} \begin{bmatrix} S_0(\omega) \\ S_0(\omega) \\ S_0(\omega) \end{bmatrix}$$

In the previous equation, the overhead tilde ($\tilde{S}$) stands for model value. Next, we compute the vector outer product as in Equation 5.66

$$\tilde{\mathbf{S}}\mathbf{S}^H = \tilde{\mathbf{\Gamma}}\mathbf{S}_0\mathbf{S}_0^H\mathbf{\Gamma}^H$$

$$= \left|\mathbf{S}_0\right|^2 \tilde{\mathbf{\Gamma}}\begin{bmatrix} 1 & 1 & 1 \\ 1 & 1 & 1 \\ 1 & 1 & 1 \end{bmatrix}\mathbf{\Gamma}^H = \left|\mathbf{S}_0\right|^2 \begin{bmatrix} \sum_{p=1}^{3}\sum_{p'=1}^{3} \tilde{a}_p^1 a_{p'}^{1^*} e^{-j(\tilde{\tau}_p^1 - \tau_{p'}^1)\omega} & & & \text{off diagonal terms} \\ & \sum_{p=1}^{3}\sum_{p'=1}^{3} \tilde{a}_p^2 a_{p'}^{2^*} e^{-j(\tilde{\tau}_p^2 - \tau_{p'}^2)\omega} & & \\ & & \ddots & \\ \text{off diagonal terms} & & & \sum_{p=1}^{3}\sum_{p'=1}^{3} \tilde{a}_p^M a_{p'}^{M^*} e^{-j(\tilde{\tau}_p^M - \tau_{p'}^M)\omega} \end{bmatrix}$$

When the test transmitter is at the correct location (i.e., $\tilde{\mathbf{r}} = \mathbf{r}$), provided the model parameters and in particular delays are correctly estimated, all diagonal terms become real and the off-diagonal terms remain complex. This fact is exploited to ascertain when a test transmitter has reached the unknown transmitter. A simulation study was carried out. We considered a shallow sea channel with five sonobuoys, each with a suspended hydrophone and a wireless transmitter to track a distant acoustic transmitter. The depth of the sea channel is 200 m. We have computed the inverse ratio (here the inverse ratio is defined as a real part divided by the imaginary part) for the test transmitter placed on the z-axis and x-axis. The computed results are shown in Tables 5.3 and 5.4. Clearly, the inverse ratio is very large at the actual location of the transmitter, which is at a depth of 37 m and range ($x = 200$, $y = 0.0$). Throughout the previous numerical experiment, the SNR (on linear scale) at sensors is maintained at 10 (on a linear scale).

TABLE 5.3
Vertical Scan through Transmitter, Located at a Range of 200 m

Depth (m)	Inverse Ratio
33	0.0250
34	20.064
35	0.7336
36	4.3366
37	**366.91**
38	4.2155
39	0.1697
40	0.7506
41	0.0360
42	1.0862

Note: The inverse ratio peaks at the correct depth, as shown by the bold row.

TABLE 5.4
Horizontal Scan through Transmitter, Located at a Range of 200 m

Range (m)	Inverse Ratio
150	1.4264
160	0.4921
170	1.1134
180	1.9129
190	0.1046
200	**940.8036**
210	0.3123
220	3.8946
230	0.2545
240	0–2096
250	2.7273

Note: The depth of the transmitter is known (at 37 m). The inverse ratio peaks at the correct range (200 m), as shown by the bold row.

The method studied earlier turns out to be sensitive to errors in sensor locations. Errors even of the order $\pm 0.1\%$ can reduce the peak height and create large side lobes, despite averaging over the entire bandwidth (3000 Hz). Results of a numerical simulation for two different sensor position errors (0.01% and 0.1%) are shown in Figure 5.16. The position error is measured with reference to the presumptive range to transmitter. A band-limited signal with a center frequency of 1500 Hz and a bandwidth of 3000 Hz is transmitted.

5.6 SUMMARY

Localization of a source becomes a simple linear problem with the combined use of Doppler shift information (FDoA) and TDoA. First, we compute two horizontal motion components and the distance to the transmitter. We use this information to compute the x- and y-coordinates of the transmitter and, finally, the z-coordinate. At least four sensors are required for the estimation of six parameters. It is also possible to compute all six parameters in one go with the help of an annihilating matrix.

A cluster of randomly distributed sensors array behaves like a linear array as it is capable of measuring DoA to one or more transmitters. It converts ToA (TDoA in case the transmitter is in the far field) information into DoA. The ML approach is followed to measure the location coordinates of all the transmitters. The DoA with reference to the center of the DSA (assumed to be at (0,0)) is easily computed. But the computer power required for non-linear optimization is beyond the scope of low-power, distributed sensors. The EM algorithm, on the other hand, is an iterative procedure for the estimation of the ML parameters of just one transmitter at a time.

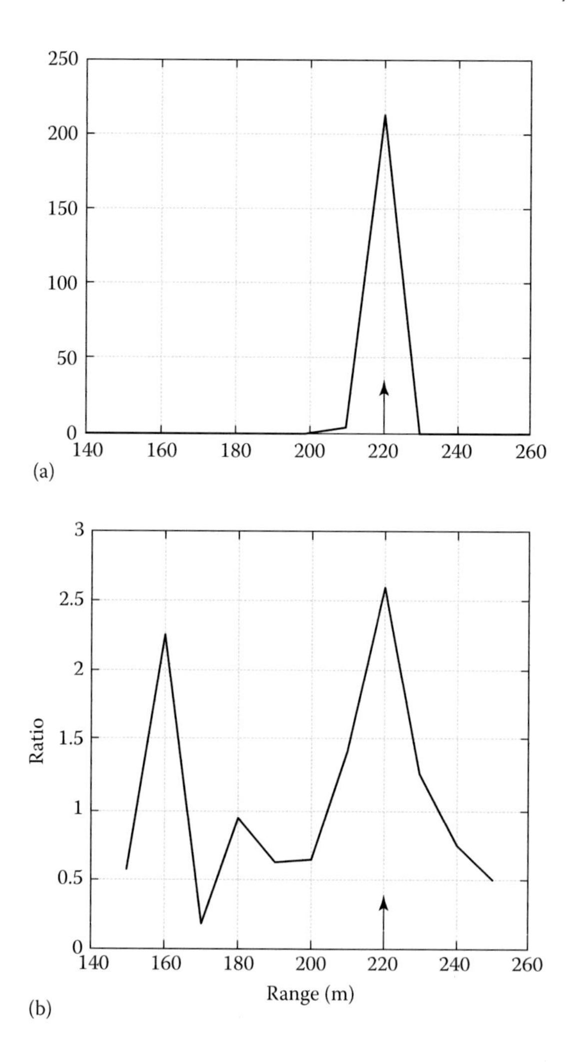

FIGURE 5.16 Effect of errors in sensor location (a) 0.01% and (b) 0.1% with reference to the presumptive transmitter location. A 64-sensor array randomly distributed inside a cuboid (±100, ±100, 0–50) m^3 was assumed. The x-coordinate of the transmitter is at 220 m (indicated by arrow).

The trick used is to consider a logarithm of the likelihood function. This enables us to express log likelihood function as a sum of the individual log likelihood functions.

When a transmitter is in a close proximity to reflecting surfaces, such as six walls of a room, two walls of a corridor, or a shallow sea channel, a DSA is required to localize the transmitter in near field. A DSA is required to localize the transmitter in the near field. In the time domain at each sensor, delays relative to the time of transmission are estimated through a cross-correlation between transmitted and received signals for different test locations of the transmitter. The true location is where the correlation is at the maximum. Theoretically, just one sensor is enough for localization when the transmitted signal is truly broadband.

For a broadband signal, we work in the frequency domain. Each received signal is Fourier transformed. We construct a column vector of large magnitude Fourier coefficients. Next, we place a transmitter at a test location and compute the theoretical column vector. For this we need to use prior knowledge of the channel model. We compute the outer product of two column vectors, which results in a matrix. At the correct location of the transmitter, this matrix will have all its diagonal terms real and the off-diagonal terms will be complex. This property has been exploited for localization; it is, however, sensitive to noise.

REFERENCES

1. A. Amar, G. Leus, and B. Friedlander, Emitter localization given time delay and frequency shift measurements, *IEEE Transactions on Aerospace and Electronics Systems*, vol. 48, pp. 1826–1836, 2012.
2. A. Weiss, Direct geolocation of wideband emitters based on delay and doppler, *IEEE Transactions on Signal Processing*, vol. 59, pp. 2513–2521, 2011.
3. A. Amar, Y. Wang, and G. Leus, Extending the classical multidimensional scaling algorithm given partial pairwise distance measurements, *IEEE Signal Processing Letters*, vol. 17, pp. 473–476, 2010.
4. A. Amar and A. J. Weiss, Direct position determination of multiple radio signals, *ICASSP*, pp. II-81–84, 2004.
5. A. Amar and A. J. Weiss, Direct position determination (DPD) of multiple known and unknown radio-frequency signals, *12th European Signal Processing Conference*, pp. 1115–1118, 2004.
6. K. C. Ho and W. Xu, An accurate algebraic solution for moving source location using TDOA and FDOA measurements, *IEEE Transactions on Signal Processing*, vol. 52, pp. 2453–2463, 2004.
7. P. S. Naidu, *Modern Spectrum Analysis of Time Series*, Boca Raton, FL: CRC Press, p. 288, 1996.
8. K. K. Mada, H.-C. Wu, and S. S. Iyengar, Efficient and robust EM algorithm for multiple wideband source localization, *IEEE Transactions on Vehicular Technology*, vol. 58, pp. 3071–3075, 2009.
9. C. E. Chen, F. Lorenzelli, R. E. Hudson, and K. Yao, Maximum likelihood DOA estimation of multiple wideband sources in the presence of non uniform sensor noise, *EURASIP Journal on Advances in Signal Processing*, vol. 2008, pp. 1–12, 2008.
10. K. K. Mada and H.-C. Wu, EM algorithm for multiple wideband source localization, *Proceedings of the IEEE Globecom. Conference*. pp. 1–5, 2006
11. L. Lu and H.-C. Wu, Novel robust direction-of-arrival-based source localization algorithm for wideband signals, *IEEE Transactions on Wireless Communications*, vol. 11, pp. 3850–3859, 2012.
12. A. P. Dempster, N. M. Laird, and D. B. Rubin, Maximum likelihood from incomplete data via the EM algorithm, *Royal Statistical Society*, *Series B*, vol. 39, pp. 1–38, 1977.
13. S. M. Jesus, M. B. Porter, Y. Stephan, X. Demoulin, O. C. Rodrigues, and E. M. M. F Coelho, Single hydrophone source localization, *IEEE Journal of Ocean Engineering*, vol. 25, pp. 337–346, 2000.
14. H. R. Uday Shankar, Broadband source localization in a wedge shaped shallow sea, PhD Thesis, Faculty of Engineering, Indian Institute of Science, Bangalore, India, 1993.
15. P. S. Naidu, *Sensor Array Signal Processing*, 2nd Edition, Boca Raton, FL: CRC Press, 2009.

6 Self-Localization

Recent advances in radio frequency signals (RF) and micro-electromechanical system (MEMS) integrated circuit (IC) design have made it possible to deploy arrays with a large number of sensors for a variety of monitoring and control applications. For example, earthquake monitoring, traffic monitoring, environment monitoring, flood monitoring, animal tracking, and so on, are some of the current application areas for sensor arrays. Automatic localization of sensors is a key enabling technology. In some applications, such as warehousing and manufacturing logistics, localization of sensors can be the main requirement. Large sensor arrays for monitoring work need to be inexpensive, energy conserving, and easily deployable even in inaccessible or hazardous environments. It is therefore not possible to precisely pre-localize sensors, for example, even after carrying out a precise engineering survey, or incorporating a modern global positioning system (GPS) receiver in each sensor. Sensors may be air-dropped or fixed on floating sonobuoys in marine applications or fired with artillery fire into enemy territory. In all such cases, the sensors will be spread over a large area in unknown locations.

6.1 ANCHOR-BASED LOCALIZATION

In anchor-based localization, the precise position of a few sensors is known beforehand, possibly using a GPS receiver. All anchor nodes are connected to a central processor, which can do number crunching for localization. Let there be M_a anchor sensors and M sensors whose locations are unknown, often called blindfolded (see Figure 6.1). Note that in the absence of anchor sensors, it is not possible to obtain, from the relative distance measurements, unambiguous self-localization. Linear translation, rotation, and mirror image cannot be corrected. To achieve absolute localization, the locations of a few (≥ 3) sensors must be known. These are called anchor sensors, optimally located on a circle enclosing all unknown sensors [1]. The anchor sensors can also be used as beacons, transmitting localizing signals to the sensors. Every sensor is then able to estimate the time of arrival (ToA) from the active anchor and relay the ToA information back to the anchor. Since anchor nodes are equipped with greater computing power, they can estimate the location of all blindfolded sensors. At least three anchor nodes must be present and within the reach of all blindfolded sensors. This is illustrated in Figure 6.1. This distributed sensor array (DSA) has three anchors and eight blindfolded sensors at unknown locations ($M = 8$); each anchor broadcasts a known signal and pre-set time of broadcast.

Blindfolded sensors receive the transmitted signal and determine the ToA. The central processor receives the ToAs from all anchors and computes the location of all blindfolded sensors.

In another possibility, the blindfolded sensors have enough power to transmit the signal and the time of broadcast to the nearest anchor, which will then estimate the

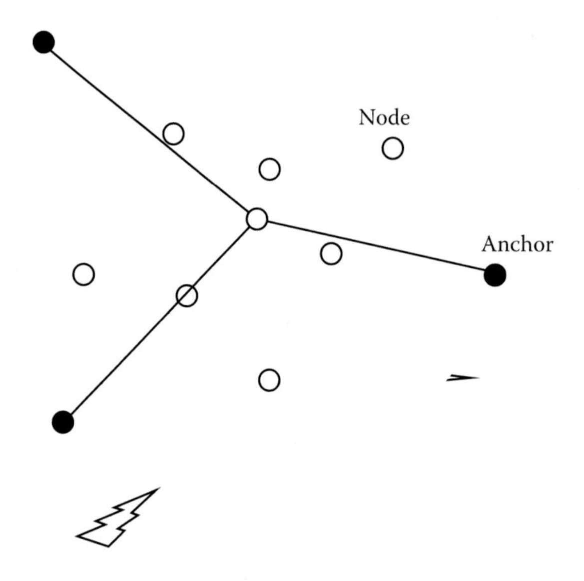

FIGURE 6.1 A DSA with three anchor sensors ($M_a = 3$).

ToA and transmit the information to the central processor. In both alternatives, it is essential that the clocks on all sensors be synchronized.

We have described two approaches (in Chapter 4) to estimate the location of an unknown source given three or more ToA measurements. Let us first consider the simpler method. In Equation 4.22a, the unknown vector $[x_{s1}\ y_{s1}]^T$ now refers to the coordinates of an unknown sensor, say, s_1. We now have M equations of the Equation 4.22b type. All these may be combined into a single composite equation, which can be solved in a single step:

$$\mathbf{A}\begin{bmatrix} x_{s_1} & x_{s_2} & \cdots & x_{sM} \\ \\ y_{s_1} & y_{s_2} & \cdots & y_{sM} \end{bmatrix} = \begin{bmatrix} \mathbf{B}_{s_1} & \mathbf{B}_{s_2} & \cdots & \mathbf{B}_{sM} \end{bmatrix} \tag{6.1}$$

where

$$\mathbf{A} = \begin{bmatrix} x_2 - x_1 & y_2 - y_1 \\ x_3 - x_1 & y_3 - y_1 \\ \cdots & \\ x_{M_a} - x_1 & y_{M_a} - y_1 \end{bmatrix}$$

and

$$\mathbf{B}_{s_m} = \frac{1}{2}\begin{bmatrix} x_2^2 + y_2^2 - & (c\tau_{2m})^2 & -x_1^2 - y_1^2 + (c\tau_{1m})^2 \\ x_3^2 + y_3^2 - & (c\tau_{3m})^2 & -x_1^2 - y_1^2 + (c\tau_{1m})^2 \\ & \cdots & \\ x_{M_a}^2 + y_{M_a}^2 & -(c\tau_{M_a m})^2 & -x_1^2 - y_1^2 + (c\tau_{1m})^2 \end{bmatrix}$$

In the previous matrix, $\tau_{1m}, \tau_{2m}, \ldots, \tau_{M_a m}$ refer to ToAs from the mth sensor to all M_a anchors. The least-squares solution of Equation 6.1 is given by

$$\begin{bmatrix} x_{s_1} & x_{s_2} & \cdots & x_{sM} \\ y_{s_1} & y_{s_2} & \cdots & y_{sM} \end{bmatrix} = \frac{1}{2}[\mathbf{A}^T\mathbf{A}]^{-1}\mathbf{A}^T\left[\mathbf{B}_{s_1}\ \mathbf{B}_{s_2}\ \cdots\ \mathbf{B}_{sM}\right] \tag{6.2}$$

EXAMPLE 6.1

Twenty sensors (blindfolded) and eight anchor nodes make up a random array as shown in Figure 6.2. The anchor nodes broadcast a known signal and time of broadcast to all sensors.

The ToAs are estimated by all sensors and relayed back to the nearest anchor node. Equation 6.2 was used to compute the sensor locations. Note that the locations of the anchor nodes are known. A small error (1%) in the ToAs was introduced to account for the errors in the anchor node location or in clock synchronization. The estimation results are shown in Table 6.1.

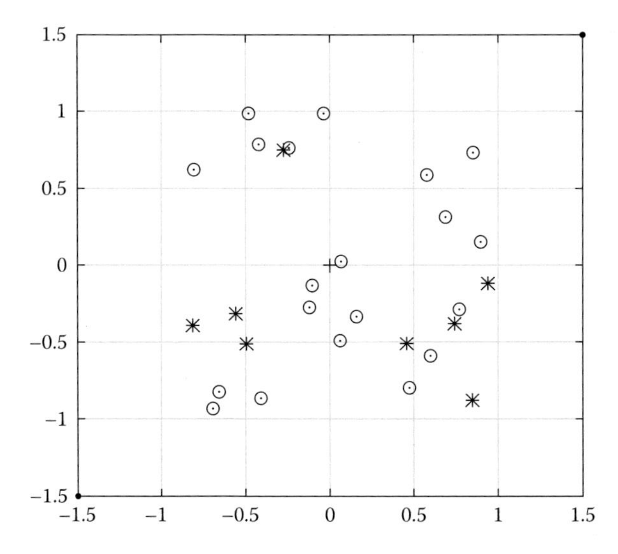

FIGURE 6.2 Randomly located DSA with 20 sensor nodes (empty circles) and 8 anchor nodes (stars).

TABLE 6.1
Actual Sensor Locations and Estimated Locations Using Equation 6.2

Actual Sensor Locations		Est. Sensor Loc. (No ToA Errors)		Est. Sensor Loc. (1% ToA Errors)	
x	y	x	y	x	y
0.1560	0.3379	0.1560	0.3379	0.1590	0.3416
0.0545	0.4928	0.0545	0.4928	0.0526	0.4913
0.4118	0.8667	0.4118	0.8667	0.4145	0.8719
0.1205	0.2747	0.1205	0.2747	0.1225	0.2840
0.5941	0.5916	0.5941	0.5916	0.5883	0.5884
0.4237	0.7834	0.4237	0.7834	0.4239	0.7940
0.4704	0.8002	0.4704	0.8002	0.4732	0.8043
0.0609	0.0215	0.0609	0.0215	0.0582	0.0201

Note: One percent error was introduced in each ToA. This results in an error in the second/third decimal places.

Given that there are just two anchors, it is not possible to uniquely localize a sensor. Clearly, there are two possible solutions, as seen in Figure 6.3. The second position is a mirror image of the first. To see this, we shall position the first anchor at (0, 0) and the second anchor at (d_{12}, 0). Then, it is easy to show that

$$d_{a_1 s}^2 = x_s^2 + y_s^2 \tag{6.3a}$$

and

$$d_{a_2 s}^2 = x_s^2 - 2x_s d_{a_1 a_2} + d_{a_1 a_2}^2 + y_s^2 \tag{6.3b}$$

Subtracting Equation 6.3a from Equation 6.3b, we obtain [2]

$$x_s = \frac{-d_{a_2 s}^2 + d_{a_1 s}^2 + d_{a_1 a_2}^2}{2d_{a_1 a_2}} \tag{6.4a}$$

and from Equation 6.3a

$$y_s = \pm\sqrt{d_{a_1 s}^2 - x_s^2} \tag{6.4b}$$

The $\pm$ sign implies that a mirror image of s_3 is also a solution for Equation 6.3. This non-uniqueness can be overcome when we consider the sensors, which are two or more hops away from the anchors.

Three anchors and two sensors within communication range can yield unique localization. Consider a situation as depicted in Figure 6.4. Sensors (s_1 and s_2) are within communication range of three anchors (a_1, a_2, and a_3). From range information alone, the first sensor has two likely locations, namely s_1 and its image s_1'.

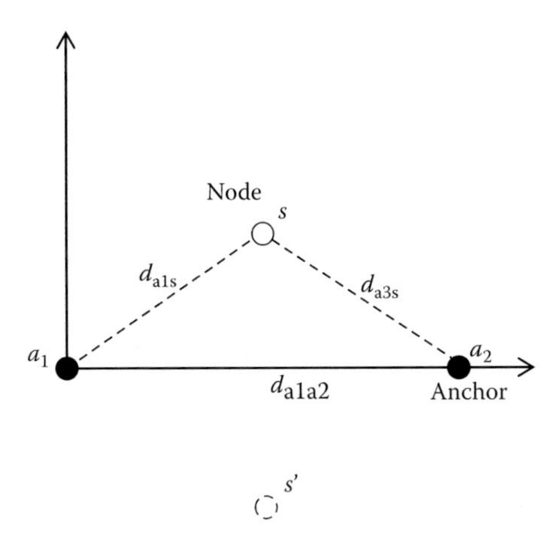

FIGURE 6.3 Two anchors (a_1 and a_2) and one node/sensor (s) does not yield unique sensor location. s', an image of s, is also a likely location of the sensor.

Similarly, the second sensor has two likely locations, namely s_2 and its image s_2'. We must now make the right choice consistent with the known distance between the two sensors. For this, we need to compute the distances between s_1 and s_2, s_1 and s_2', s_2 and s_1', and s_1' and s_2' (four distances). Next, we compare the computed distances with the known distance. The closest match decides the sensor pair as the correct answer.

The method will give the correct answer as long as both sensors are inside or outside the triangle formed by the three anchors. Ambiguity, however, can arise whenever one sensor is inside and the other is outside but symmetrically placed in a triangle with at least two equal sides. One such possibility is illustrated in Figure 6.5. The distance criterion, as used in Figure 6.4, is no longer valid as the distance between sensors and their images is equal. However, for oblique triangles the distance criterion is still useful to overcome the ambiguity.

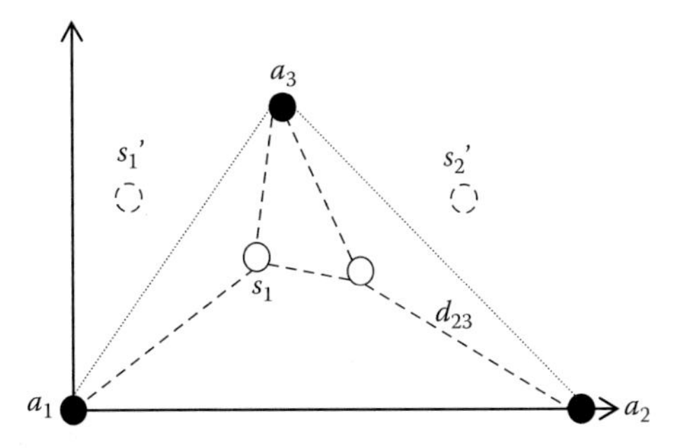

FIGURE 6.4 Two sensors (s_1 and s_2) are within the communication range of three anchors (a_1, a_2, and a_3) as shown. But no sensor lies within the communication range of all three anchors. Both sensors are within communication range of each other. Compute distances between four pairs of sensor nodes. The closest match decides the sensor pair as the correct answer.

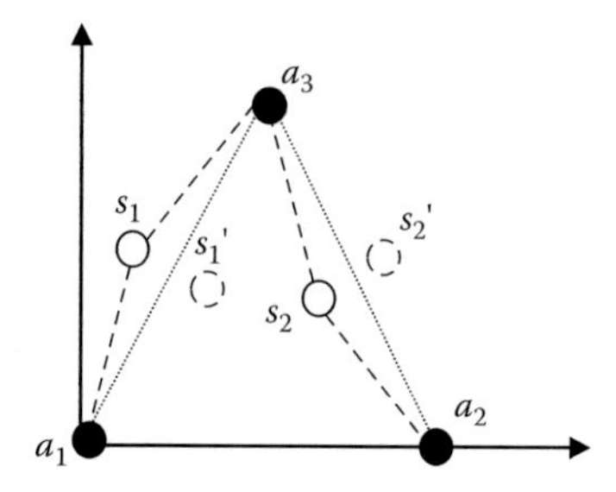

FIGURE 6.5 In a bilateral triangle, sensors s_1 and s_2 and their images s_1' and s_2' are both likely locations.

6.1.1 Cooperative Localization

A large sensor network, having a few hundred to a few thousand sensors spread over a large area in some arbitrary form, is a challenging problem in DSA. Sensors are not wired and there is no direct radio link to a central processor; they are able to communicate only with their immediate neighbors. To maintain low power consumption and low cost, the sensors will have very limited on-board processing capability. Each sensor can communicate with its neighbors, which are within communication range. It can also estimate the relative distance to its neighbors. As an illustration, in Figure 6.6, we have sketched a DSA showing 18 sensors, which are able to communicate and measure their relative distances. In cooperative localization, we use such measured relative distances to determine the shape of the array. The shape can be established

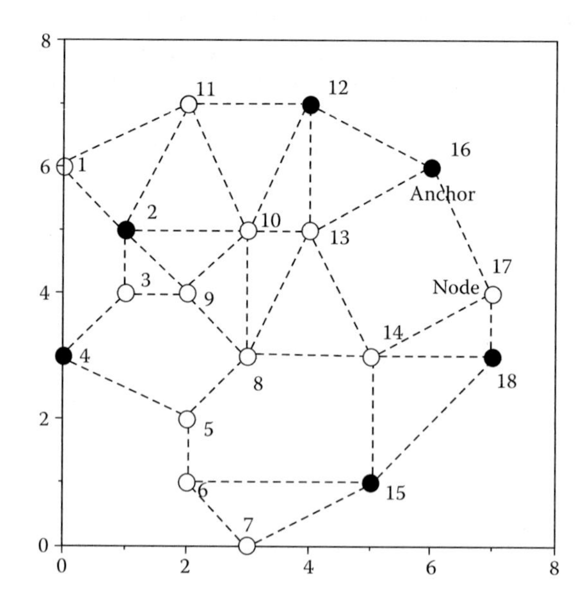

FIGURE 6.6 Every node/sensor can communicate with its neighbors, which lie within its communication range. It is also able to measure its relative distance. Dashed lines show nodes within the communication range of an anchor.

but for a few ambiguities, namely, translation, rotation, and mirror image, which may be overcome provided there are three or more anchor sensors. Because of the power restriction, the distance between the sensors is a few tens of meters. We shall assume that pairwise distance estimates have been carried out.

One strategy is to look for a possibility where a sensor is within one hop from at least three anchors. Then, it is possible to localize the sensor, which can then be treated as an additional anchor. We have six anchors. But there is no sensor within one hop away from the three anchors. However, there is another interesting situation, where two sensors (14 and 17) are within communication reach of three anchor sensors. To emphasize this, we have redrawn the network in Figure 6.6, leaving out all non-relevant connections (see Figure 6.7). Sensors 14 and 17 are within one hop of the three anchors and also within themselves. But each sensor communicates with only two out of the three anchors. The sensors lie outside a triangle formed by the three anchors. Hence, we explore the possibility of applying the method illustrated in Figure 6.4. We shall now show how to localize the unknown sensors in this situation. Since sensor 14 can communicate with anchors 15 and 18, we can localize sensor 14 and its image. Similarly, sensor 17 communicates with anchors 16 and 18; we can localize sensor 17 and its image. Compute distances between sensors 14 and 17, between their images, between sensor 14 and the image of 17, and between sensor 17 and the image of 14. We need to choose one of the four possibilities by comparing the known distance between sensors 14 and 17. We choose the one giving a close match. Once a sensor is located, it may be treated as an anchor and is then used for localizing other unknown sensors. For example, in Figure 6.8, treating sensors 14 and 17 as anchor nodes, it is possible to localize sensor 13. Finally, treating sensor 13 as an anchor along with anchors 2 and 12, we can localize sensor 10. The previously mentioned strategy can be extended to a complex arbitrary DSA, but it may become very laborious and sometimes it may even fail to cover the entire network.

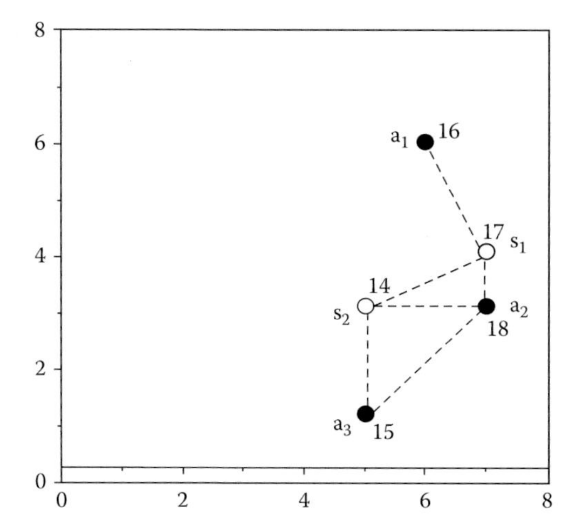

FIGURE 6.7 Two blindfolded sensors (14 and 17) are within communication range of three anchor sensors (15, 16, and 18), shown with dashed lines, and also within communication range of each other.

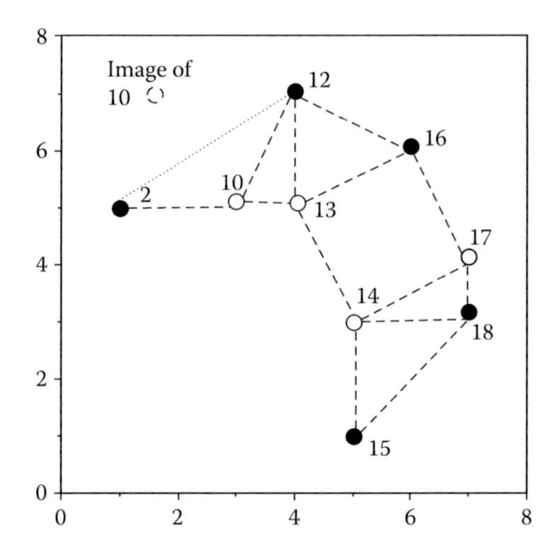

FIGURE 6.8 Given two anchors, localization is non-unique. There is a valid mirror image of the sensor that is being localized. The mirror image of sensor 10 across the line joining anchors 2 and 12 does not satisfy the distance constraint with sensor 13, whose position has already been fixed.

6.1.2 Subspace Positioning Approach

The subspace algorithm used in direction of arrival (DoA) (also known as MUSIC [3]) has been shown to be applicable in sensor node localization [4]. The basic information used is the distance-squared matrix **D**, whose (ith and jth) element is the squared distance between the ith and the jth sensors. Matrix **D** is symmetric and diagonal; its entries are zeros. The sensor array may have any random shape and distribution, but all internode distances must be estimated using either ToA/ time difference of arrival (TDoA) or relative signal strength (RSS). The range-based matrix contains a significant amount of information on the sensor location. In the next section, we describe an algorithm known as multidimensional scaling to compute the relative coordinates of all the sensors from the **D** matrix. Here, we shall describe a method of computing the universal coordinates of all but a few sensors. We must have the absolute coordinates of three or more sensors, which are known as anchors, probably equipped with GPS devices, extra computing power, and large battery power. We briefly outline Chan's distributed subspace algorithm, which is more appropriate for our application, where we have no central processing unit.

Consider an array of randomly distributed M sensors, out of which there are $M_a \geq 3$ anchors. Define the matrix anchor coordinates:

$$\mathbf{X}_m = \begin{bmatrix} x_m - x_1 & x_m - x_2 & \cdots & x_m - x_{M_a} \\ y_m - y_1 & y_m - y_2 & \cdots & y_m - y_{M_a} \end{bmatrix}^T$$
$$= \begin{bmatrix} \mathbf{I}_{M_a} & \mathbf{0}_{M_a \times 1} \end{bmatrix} x_m + \begin{bmatrix} \mathbf{0}_{M_a \times 1} & \mathbf{I}_{M_a} \end{bmatrix} y_m - \begin{bmatrix} \mathbf{x}_{M_a} & \mathbf{y}_{M_a} \end{bmatrix} \tag{6.5}$$

where $(x_1, x_2, \cdots x_{M_a}$ and $y_1, y_2, \cdots y_{M_a})$ are x- and y-coordinates of the anchors and (x_m, y_m) is the coordinate of the mth unknown sensor. Vectors $\mathbf{x}_{M_a}$ and $\mathbf{y}_{M_a}$ are defined as

$$\mathbf{x}_{M_a} = \begin{bmatrix} x_1 & x_2 & \cdots & x_{M_a} \end{bmatrix}^T$$

$$\mathbf{y}_{M_a} = \begin{bmatrix} y_1 & y_2 & \cdots & y_{M_a} \end{bmatrix}^T$$

Construct another matrix $\mathbf{F}_m$ using $\mathbf{X}_m$ as follows:

$$\begin{aligned} \mathbf{F}_m &= \mathbf{X}_m \begin{bmatrix} \mathbf{X}_{M_a+1}^T & \mathbf{X}_{M_a+2}^T & \cdots & \mathbf{X}_M^T \end{bmatrix} \\ &= -0.5\mathbf{P}_m \mathbf{H}_m^T \mathbf{D} \begin{bmatrix} \mathbf{H}_{M_a+1}\mathbf{P}_{M_a+1}^T & \mathbf{H}_{M_a+2}\mathbf{P}_{M_a+2}^T & \cdots & \mathbf{H}_M \mathbf{P}_M^T \end{bmatrix} \end{aligned} \tag{6.6a}$$

where

$$\mathbf{P}_m = \begin{bmatrix} \mathbf{I}_{M_a} & \mathbf{0}_{M_a \times (m-1-M_a)} \end{bmatrix}$$

and

$$\mathbf{H}_m = \begin{bmatrix} -\mathbf{I}_{m-1} & \mathbf{1}_{m-1} & \mathbf{0}_{m-1 \times (M-m)} \end{bmatrix}^T,$$

$$m = M_a + 1, M_a + 2, \cdots, M$$

The size of $\mathbf{F}_m$ is $M_a \times (M - M_a)$, but its rank is two. Construct a square matrix (size: $M_a \times M_a$)

$$\mathbf{FF}_m = \mathbf{F}_m \mathbf{F}_m^T \tag{6.6b}$$

The eigen decomposition of Equation 6.6b may be expressed as $\mathbf{FF}_m = \mathbf{U}\lambda\mathbf{U}^T$ where the eigenvalue matrix λ has only two non-zero (positive) eigenvalues.

$$\boldsymbol{\lambda} = \begin{bmatrix} \lambda_1 & 0 & 0 \cdots 0 \\ 0 & \lambda_2 & 0 \cdots 0 \\ 0 & 0 & 0 & 0 \\ \vdots & & \vdots & \vdots \\ 0 & 0 & 0 \cdots 0 \end{bmatrix}_{M_a \times M_a}$$

Let $\mathbf{u}_\eta$ be the eigenvector corresponding to zero eigenvalue(s). Pre- and post-multiply $\mathbf{FF}_m$ with $\mathbf{X}_m$; we get the basic result

$$\mathbf{u}_\eta^T \mathbf{X}_m = \mathbf{0}_{(M_a-2)\times 2} \tag{6.7}$$

which can further be simplified using the expansion of $\mathbf{X}_m$ from Equation 6.5 and using a "vec" operation (Octave command).

$$\left[vec\left(\mathbf{u}_\eta^T\left[\mathbf{1}_{M_a} \quad \mathbf{0}_{M_a\times 1}\right]\right) \quad vec\left(\mathbf{u}_\eta^T\left[\mathbf{1}_{M_a} \quad \mathbf{0}_{M_a\times 1}\right]\right)\right]\begin{bmatrix} x_m \\ y_m \end{bmatrix} = vec\left(\mathbf{u}_\eta^T\left[\mathbf{x}_{M_a} \quad \mathbf{y}_{M_a}\right]\right) \tag{6.8a}$$

The least-squares solution of Equation 6.8a is given by

$$\begin{bmatrix} x_m \\ y_m \end{bmatrix} = \mathbf{A}_m^{-1}\mathbf{B}_m \tag{6.8b}$$

where

$$\mathbf{A}_m = \left[vec\left(\mathbf{u}_\eta^T\left[\mathbf{1}_{M_a} \quad \mathbf{0}_{M_a\times 1}\right]\right) \quad vec\left(\mathbf{u}_\eta^T\left[\mathbf{1}_{M_a} \quad \mathbf{0}_{M_a\times 1}\right]\right)\right]_{2\times 2}$$

$$\mathbf{B}_m = vec\left(\mathbf{u}_\eta^T[\mathbf{x}_{M_a} \quad \mathbf{y}_{M_a}]\right)$$

$\mathbf{A}_m$ is a diagonal matrix ∀, hence it is trivial to invert it. The eigen decomposition of Equation 6.6b requires a rather involved computational hardware. This may be partially overcome by using an algorithm, which iteratively finds the eigenvector corresponding to the largest eigenvalue. We can invert the matrix in Equation 6.6b so that the eigenvector corresponding to the largest eigenvalue now corresponds to that of the smallest eigenvalue.

EXAMPLE 6.2

We consider an example of a random DSA with 4 anchors and 12 sensors at random locations over a square (10*10 m^2) area as shown in Figure 6.9. The distance-squared matrix between all pairs of sensors is assumed to have been measured through either ToA or RSS measurements. Initially, to check the Octave program, we assumed no measurement errors and compared the computed and actual sensor coordinates. The match was found to be perfect. With the addition of a small measurement error, epsi, which is assumed to be a zero mean and unit variance Gaussian random variable, the error in coordinate estimation (rms error) increases rapidly, as shown in Figure 6.10. The error depends upon the specific layout of the anchors and sensor nodes.

What is shown in Figure 6.10 is for a specific layout shown in Figure 6.9. Chan [4] presented an extensive analysis of the mean square error and bias on sensor position errors.

Since the anchor sensor position is according to the users' convenience, it is useful to know how one may arrange the anchors for minimum position errors. We shall

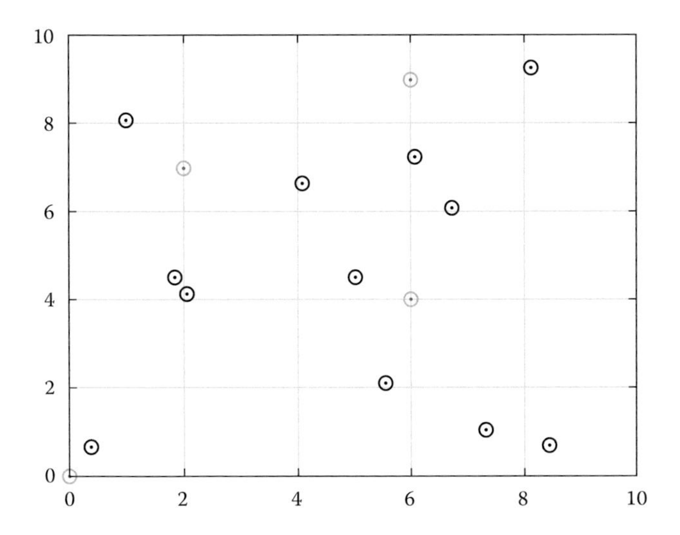

FIGURE 6.9 Sixteen sensor random DSA with 4 anchor sensors (shown as gray circles) and 12 sensors (shown as black circles). Coordinates shown are in meters.

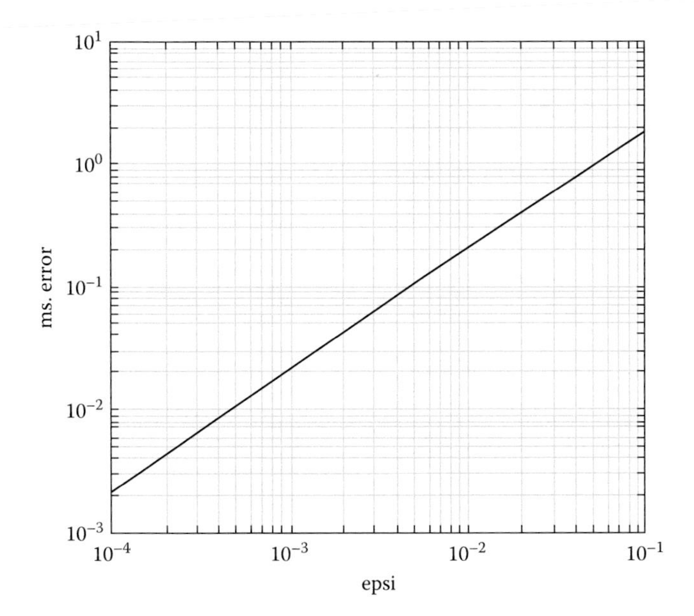

FIGURE 6.10 The root mean square error as a function of measurement error of magnitude = 2epsi. Factor two arises because of the squaring of the distance, $(d_{m,n})^2 = c^2(\tau_{m,n} + epsi)^2 \approx c^2(\tau_{m,n}^2 + 2epsi * \tau_{m,n})$. The rms error increases exponentially with epsi.

consider one particular arrangement, namely, boundary array. Four anchors are placed as shown in Figure 6.11. The rms error is much lower when the anchors are at four corners. Further reduction is possible when the anchors are mingled with sensors, as in Figure 6.9; then, the average distance is reduced, hence, lower measurement errors.

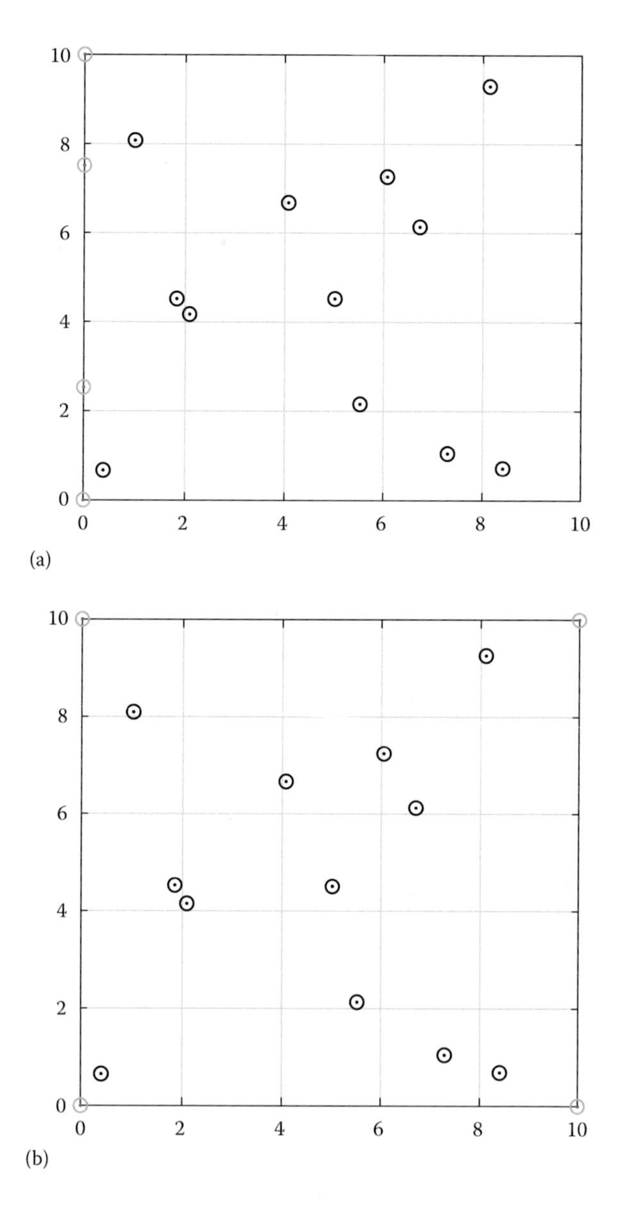

FIGURE 6.11 Twelve sensors (black circles) and 4 anchors (gray circles) placed on the boundary enclosing all sensors. The rms position errors are as follows: (a) 8.3875 (anchors, all on left side), (b) 0.4549 (anchors at four corners).

6.2 ANCHOR-FREE LOCALIZATION

Until now, the localization of sensors in a DSA depended on knowledge of the location of anchor nodes. In the absence of anchors, it is still possible to map sensor distribution, with some limitations. However, the presence of anchor nodes allows us to overcome these limitations. Also, we may have other types of information, such as the presence of some landmarks associated with the sensors; for example, a few sensors with known identifications (IDs) being placed along a road. There are other situations in which the exact location of sensors is not required. Suppose we need to broadcast a message over a large area but with limited range. With a transceiver at the base station, the message will be received by all transceivers lying within broadcast range; they in turn will rebroadcast the message to other transceivers. In no time, the whole area will be flooded with the message. The transceiver at the base station is the lone anchor required in this operation. The mapping of sensors without anchors is achieved through an algorithm called multidimensional scaling (MDS), first suggested by Torgerson [5] in 1952 and used by Kruskal [6] in 1964.

MDS is used to get the geometrical representation of objects in space based on the similarity/dissimilarity of the objects. A score (on a scale of 0–1) is attached to each object as a measure of similarity. The mean square difference is between observed similarity and reproduced distance, that is, the distance between objects in multidimensional space of given dimensions (or attributes). The sum of the mean-squared difference between all pairs of objects, also known as stress, is minimized. When the minimum is achieved, all similar objects form a cluster or group. By increasing the dimensions of the space, the stress can be further reduced; but beyond a certain number, the stress function decreases very slowly. This number is the effective dimension of the space. MDS has many applications in economics, political science, and behavioral sciences [7].

6.2.1 Multidimensional Scaling

MDS is an algorithm that constructs a map of objects (in our case sensors) by using distances as dissimilarity between sensor nodes. Given a DSA with pairwise distances, we can apply the MDS algorithm to construct a map of the array but with ambiguity of translation, rotation, and mirror image. If we have at least three anchor sensors, it is possible to remove these ambiguities.

Let M sensors be located at points p_i, $i=1,2,\ldots M$ where p_i is the coordinate vector, $p_i=(x_i,y_i)$. The information available is the distance between all possible pairs of sensors, assuming for the time being they are all within the communication range. Consider a pair of sensors i and j in 2D space and the distance between them is D_{ij}. Then

$$D_{i,j}^2 = \|\mathbf{p}_i - \mathbf{p}_j\|^2 = \left[\left(x_i - x_j\right)^2 + \left(y_i - y_j\right)^2\right] \tag{6.9a}$$

$$= (x_i^2 + y_i^2) + (x_j^2 + y_j^2) - 2(x_i x_j + y_i y_j)$$

We make an important assumption that the center of the coordinates is at the geometrical center of the sensor array, which implies that

$$\sum_i x_i = \sum_i y_i = 0 \tag{6.9b}$$

Consider the following relations

$$\bar{D}_i^{row} = \sum_j D_{i,j}^2 = Mx_i^2 + My_i^2 + \sum_j \left(x_j^2 + y_j^2\right)$$

$$\bar{D}_j^{col} = \sum_i D_{i,j}^2 = Mx_j^2 + My_j^2 + \sum_i \left(x_i^2 + y_i^2\right) \tag{6.9c}$$

$$\bar{D}_0 = \sum_i \sum_j D_{i,j}^2 = M\sum_i \left(x_i^2 + y_i^2\right) + M\sum_j \left(x_j^2 + y_j^2\right)$$

Note that the cross-terms in Equation 6.9a vanish because of Equation 6.9b. From Equation 6.9c, it is easy to show that

$$\frac{\bar{D}_i^{row}}{M} + \frac{\bar{D}_j^{col}}{M} - \frac{\bar{D}_0}{M^2} = \left(x_i^2 + y_i^2\right) + \left(x_j^2 + y_j^2\right) \tag{6.9d}$$

Using Equation 6.9d in Equation 6.9a, we obtain an interesting relation, which forms the basis of the MDS algorithm,

$$D_{i,j}^2 - \frac{\bar{D}_i^{row}}{M} - \frac{\bar{D}_j^{col}}{M} + \frac{\bar{D}_0}{M^2} = -2(x_i x_j + y_i y_j) \tag{6.10}$$

Define

$$B_{i,j} = -\frac{1}{2}\left[D_{i,j}^2 - \frac{\bar{D}_i^{row}}{M} - \frac{\bar{D}_j^{col}}{M} + \frac{\bar{D}_0}{M^2}\right] = \left(x_i x_j + y_i y_j\right) \tag{6.11a}$$

The algebraic manipulations indicated in Equations 6.9c–6.10 may be compactly represented by matrix operation

$$\mathbf{B} = -\frac{1}{2}\mathbf{JDJ} \tag{6.11b}$$

where $\mathbf{J}$ is given by

$$\mathbf{J} = \left(\mathbf{I}_{M\times M} - \frac{1}{M}\mathbf{1}_{M\times 1}\mathbf{1}_{M\times 1}^T\right)$$

where $\mathbf{I}_{M\times M}$ is an identity matrix of size ($M\times M$) and $\mathbf{1}_{M\times 1}$ is a column matrix of ones. $\mathbf{D}$ is a distance-squared matrix, that is, $\{\mathbf{D}\}_{ij} = D_{ij}^2$ and $\{\mathbf{B}\}_{ij} = B_{ij}$

Notice that $\mathbf{B}$ is a sum of the outer products of two vectors,

$$\mathbf{x} = \{x_1, x_2, ..., x_M\}$$

$$\mathbf{y} = \{y_1, y_2, ..., y_M\}$$

That is,

$$\mathbf{B} = \mathbf{x}\mathbf{x}^T + \mathbf{y}\mathbf{y}^T \tag{6.12}$$

This is a useful result, as the rank of $\mathbf{B}$, in a noise-free case, will be equal to or less than two. The latter situation will arise when all sensor nodes are clustered or lie on a line. Barring the previous degenerate cases, $\mathbf{B}$ will have two non-zero eigenvalues $\{\lambda_1,\lambda_2\}$ and the corresponding eigenvectors $\{\mathbf{u}_1,\mathbf{u}_2\}$ will be related to the position vectors x and y. In fact, the relationship is given by

$$\mathbf{x} = \mathbf{u}_1\sqrt{\lambda_1}, \quad \mathbf{y} = \mathbf{u}_2\sqrt{\lambda_2} \tag{6.13}$$

Note that the $\mathbf{B}$ in Equation 6.12 is a symmetric matrix and hence its eigenvalues are real and positive. But the eigenvectors are not unique. The relative magnitude of the eigenvalues will depend upon the spread of the sensors in the x and y directions. For example, the eigenvalues will be equal for an array that is uniformly spread over a square. Additionally, Equation 6.13 requires that the position vectors be orthogonal, that is, $\mathbf{x}^T\mathbf{y} = 0$. The sequential order in which the nodes are arranged will remain unchanged in the eigenvectors.

The MDS algorithm yields a map where the nodes suffer from an arbitrary translation, scale, rotation, and reflection. We need at least three anchor nodes (in a 2D case) in order to correct such a map. Mapped nodes corresponding to the known anchors may be identified based on their serial order by comparing the list of nodes and their map. Let one of the known anchors be positioned at the center of coordinates, (0, 0). The entire map is then laterally shifted so that the map of the node is positioned at (0, 0). Next, we correct for scale, rotation, and mirror imaging. The second anchor is placed on the x-axis keeping the distance between the two anchors unchanged. The entire map is rotated until the map of the second anchor is brought onto the x-axis. The scale factor is now estimated by comparing the distance to map of the second anchor with the actual distance of the second anchor, also on the x-axis. The rotation may be realized through matrix multiplication (Equation 6.14). Let θ be the angle of rotation, which is known from the map of the second anchor as it is dragged to the x-axis. The required rotation matrix is given by

$$\mathbf{r}(\theta) = \begin{bmatrix} \cos(\theta) & -\sin(\theta) \\ \sin(\theta) & \cos(\theta) \end{bmatrix}$$

Rotated node coordinates $[\mathbf{x}' \ \ \mathbf{y}']^T$ are obtained through a simple matrix multiplication,

$$\begin{bmatrix} \mathbf{x}' \\ \\ \mathbf{y}' \end{bmatrix} = \mathbf{r}(\theta) \begin{bmatrix} \mathbf{x} \\ \\ \mathbf{y} \end{bmatrix} \tag{6.14}$$

Note that $\mathbf{r}(\theta)$ is a unitary matrix; therefore, its inverse is simply its transpose.

Finally, the third anchor is used to decide upon possible reflection, if any. The reflection is examined by comparing the third anchor and its map. When these two are at the same point, we may conclude that there is no reflection. Reflection may be realized again through a simple matrix multiplication:

$$\begin{bmatrix} 1 & 0 \\ 0 & 1 \end{bmatrix} \quad \text{No reflection}$$

$$\begin{bmatrix} 1 & 0 \\ 0 & -1 \end{bmatrix} \quad \text{Reflection across x-axis}$$

$$\begin{bmatrix} -1 & 0 \\ 0 & 1 \end{bmatrix} \quad \text{Reflection across y-axis}$$

Through a simple examination of the x- and y-coordinates of the third anchor and its image, we can choose one of the previous reflections. But it is more difficult when the reflection is across a line making an arbitrary angle φ. To get the image position of the third anchor, we drop a perpendicular to a line making an angle φ. The equation of the line is simply $\sin\varphi\, x - \cos\varphi\, y = 0$ and the foot of the perpendicular is

$$\begin{aligned} x_2 - x_1 &= \sin\varphi\,(-\sin\varphi\, x_1 + \cos\varphi\, x_2) \\ y_2 - y_1 &= \cos\varphi\,(\sin\varphi\, y_1 - \cos\varphi\, y_2) \end{aligned} \tag{6.15a}$$

where (x_1,y_1) refer to anchor coordinates and (x_2,y_2) refer to that of the foot of the perpendicular. Extending the perpendicular further down the same length yields image point. Its coordinates (x_3,y_3) are given by

$$\begin{aligned} x_3 &= x_2 + (x_2 - x_1) \\ y_3 &= y_2 - (y_2 - y_1) \end{aligned} \tag{6.15b}$$

Eliminating (x_2, y_2) from Equations 6.15a and 6.15b we obtain

$$x_3 = \cos(2\varphi)\, x_1 + \sin(2\varphi)\, y_1$$

$$y_3 = \sin(2\varphi)\, x_1 - \cos(2\varphi)\, y_1$$

Thus, the anchor coordinates are transformed to image coordinates by [8]

$$\boldsymbol{\rho}(\varphi) = \begin{bmatrix} \cos(2\varphi) & \sin(2\varphi) \\ \sin(2\varphi) & -\cos(2\varphi) \end{bmatrix} \tag{6.16}$$

Note that $\boldsymbol{\rho}(\varphi)$ is a unitary matrix; therefore, its inverse is its own transpose. Both operations, rotation and reflection, can be combined into a single correction matrix:

$$\boldsymbol{\rho}_{Corr} = \begin{bmatrix} \cos(\theta) & -\sin(\theta) \\ \sin(\theta) & \cos(\theta) \end{bmatrix} \begin{bmatrix} \cos(2\varphi) & \sin(2\varphi) \\ \sin(2\varphi) & -\cos(2\varphi) \end{bmatrix} \tag{6.17}$$

There is an alternative method for computing ρ_{Corr}. We select the anchor node and their images. They are related through Equation 6.18, whose least-squares solution is given by Equation 6.19a. We need at least two anchors, not collinear with the first anchor, to estimate ρ_{Corr}. Let ξ_1 and ξ_2 be the position vectors of two anchors, and the corresponding position vectors of the images are ξ_1' ξ_2'. The position vectors of anchors and their images are governed by

$$\left[\xi_1' \; \xi_2'\right] = \boldsymbol{\rho}_{corr}\left[\xi_1 \; \xi_2\right] \tag{6.18}$$

The least-squares solution of Equation 3.14 leads to

$$\rho_{corr} = \sum_{i=1}^{2} \xi_i' \xi_i^T \left(\sum_{i=1}^{2} \xi_i \xi_i^T \right)^{-1} \tag{6.19a}$$

Now, consider a situation where there are more than three anchors, say, $M+1$ anchors ($M \geq 2$). After correcting for translation, we are left with M anchors for rotation and reflection correction. The least-squares solution of the desired correction matrix is simply the generalization of Equation 6.19a

$$\rho_{corr} = \sum_{i=1}^{M} \xi_i' \xi_i^T \left(\sum_{i=1}^{M} \xi_i \xi_i^T \right)^{-1} \tag{6.19b}$$

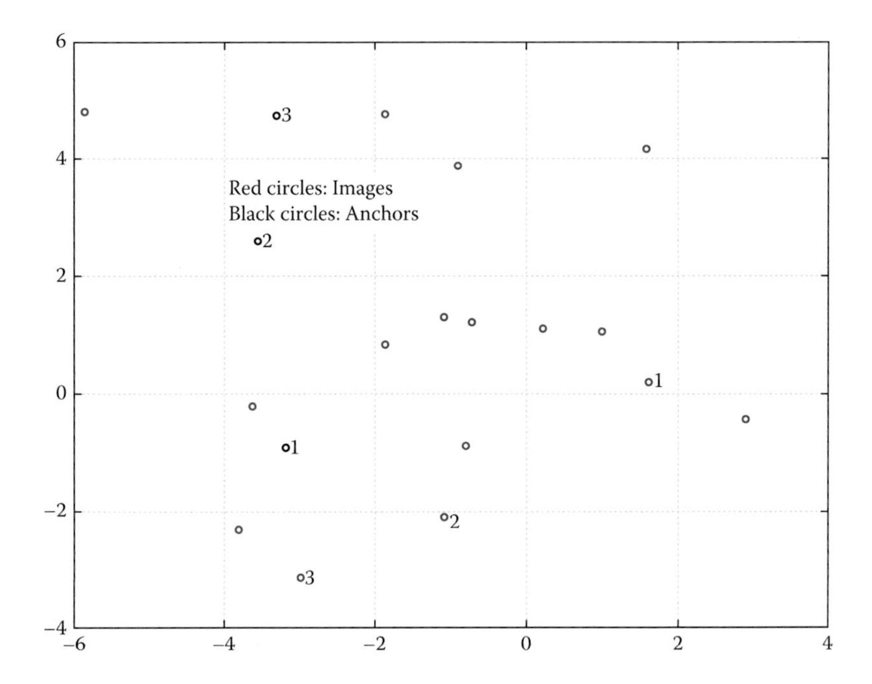

FIGURE 6.12 Result of MDS reconstruction. The positions (in m) of 16 sensors are shown as red circles, while the positions of 3 known anchors are shown as black circles.

EXAMPLE 6.3

The following is a numerical example of MDS reconstruction using range information only. There are three anchors whose locations are known and their corresponding images. In Figure 6.12, all red circles are images of sensors and black circles are anchors. The anchors and their corresponding images are numbered. Using this information, we have used Equation 6.19b to compute the correction matrix. The correction matrix applies to all image sensors, which may be used to map all images back into their actual positions. The results are shown in Table 6.2. For some sensors, their actual position and their image positions are shown in columns two and three respectively. Using the correction matrix, the image positions were mapped back into their actual positions, as shown in column four. As there was no noise, the mapping was exact without any error:

$$\rho_{Corr} = \begin{bmatrix} -0.57013 & -0.82155 \\ 0.82155 & -0.57013 \end{bmatrix}$$

6.2.2 Basis Point MDS

The B matrix in Equation 6.7b is of rank two (rank three in 3D space), but the matrix size is M, the number of sensors; hence, there is a large redundancy. This has been exploited to reduce the required inter-sensor distances M (M−1)/2. All entries in the

TABLE 6.2
Result of Mapping Images into Their Actual Position

Sensor #	Position	Image	Calculated Position
14	(2.432, 1.579)	(−2.684, 1.098)	(2.433, 1.579)
10	(−3.271, −2.235)	(−0.287, 3.961)	(3.271, −2.235)
6	(8.026, 3.514)	(−7.463, 4.591)	(8.026, 3.5143)
4	(1.057, 0.007)	(−0.608, 0.864)	(1.057, 0.007)

Note: For the sake of illustration, just three sensors are considered.

matrix D (Equation 6.9a) must be filled. It may be noted that 1D space (line) requires just two points, 2D space (plane) requires just three points, and 3D space (volume) requires just four points for their complete definition; that is, any point in the space can be expressed in terms of the coordinates of these points. These are the basis points [9]. Following the work of [10], we show how this can be achieved.

6.2.3 1D Space

We start with the 1D case: Consider two points, δ_A and δ_B, separated by δ_{AB}. The coordinate of the first point is $\delta_{AB}/2$ and that of the second point is $-\delta_{AB}/2$. Consider a point Q on the line at a distance δ_{AQ} from δ_A, and δ_{BQ} from δ_B (see Figure 6.13). Given this much information, we like to compute the x-coordinate of Q:

$$\delta_{AQ}^2 = \left(\frac{-\delta_{AB}}{2} - x_Q\right)^2 + y_Q^2 + z_Q^2 \tag{6.20a}$$

$$\delta_{BQ}^2 = \left(\frac{\delta_{AB}}{2} - x_Q\right)^2 + y_Q^2 + z_Q^2 \tag{6.20b}$$

where $(y_Q, z_Q) = 0$ are the y- and z-coordinates of point Q. Subtracting Equation 6.20b from Equation 6.20a, we obtain

$$x_Q = \frac{\delta_{AQ}^2 - \delta_{BQ}^2}{2\delta_{AB}} \tag{6.21}$$

6.2.4 2D Space

Next, we consider 2D space (plane). In Figure 6.13, we move point Q away from line AB, keeping x_Q fixed as given by Equation 6.21. Alternatively, we can select any point Q, but its distance to points A and B must be known. Next, we need to

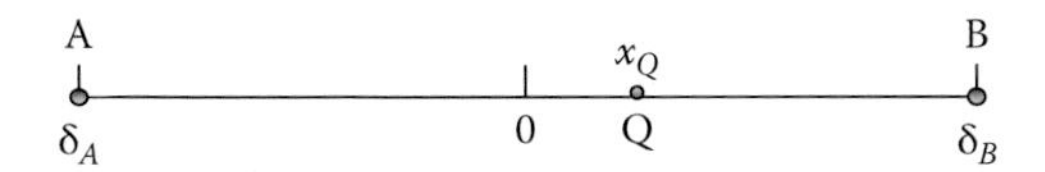

FIGURE 6.13 Q is any point on line AB (in 1D space). There are two points, δ_A and δ_B, equidistant from the center. Given distances δ_{AQ} and δ_{BQ}, we can compute x_Q, x-coordinate of Q.

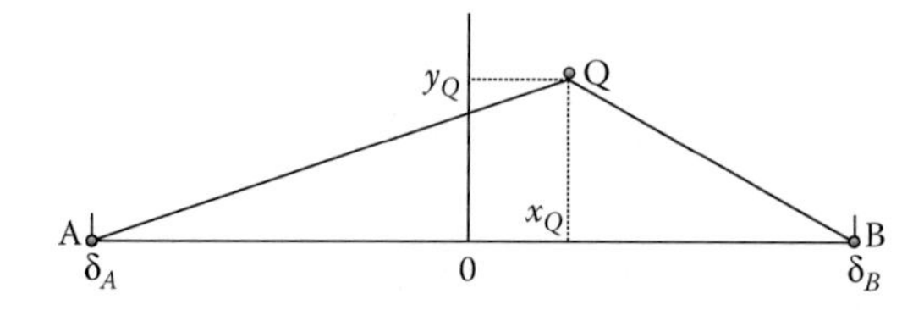

FIGURE 6.14 Q is a third point. A plane drawn through these three points. δ_{AQ} and δB_Q are known. Further, given the coordinates of point A and point x_Q, we can compute the y-coordinate of point Q (see Equation 6.22).

compute y_Q ($z_Q=0$); given AQ, BQ, and AB, we can easily show that x_Q is given by Equation 6.21. In Figure 6.14 in $\Delta A x_Q Q$

$$y_Q^2 = \delta_{AQ}^2 - \frac{\delta_{AB}^2}{4} - x_Q^2 - \delta_{AB} x_Q \tag{6.22a}$$

and in $\Delta B x_Q Q$

$$y_Q^2 = \delta_{BQ}^2 - \frac{\delta_{AB}^2}{4} - x_Q^2 + \delta_{AB} x_Q \tag{6.22b}$$

Adding Equations 6.22a and 6.22b, we obtain

$$y_Q = \sqrt{\frac{\delta_{AQ}^2}{2} - \frac{\delta_{AB}^2}{4} + \frac{\delta_{BQ}^2}{2} - x_Q^2} \tag{6.22c}$$

or

$$x_Q^2 + y_Q^2 = \frac{\delta_{AQ}^2}{2} - \frac{\delta_{AB}^2}{4} + \frac{\delta_{BQ}^2}{2}$$

We thus obtain y_Q in terms δ_{AQ}^2, δ_{BQ}^2, and δ_{AB}^2. We have been able to obtain the x- and y-coordinates from distances QA, QB, and AB.

Let us relabel point Q as C, and the points A, B, and C will be the basis points of the plane. The coordinates of the basis points are

$$A: \ [-\delta_{AB/2}, 0]$$

$$B: \ [\delta_{AB/2}, 0]$$

$$\left[C: \ \frac{\delta_{AC}^2 - \delta_{BC}^2}{2\delta_{AB}}, \sqrt{\frac{\delta_{AC}^2}{2} - \frac{\delta_{AB}^2}{4} + \frac{\delta_{BC}^2}{2} - x_C^2} \right] \tag{6.23}$$

Let Q be any arbitrary point in the plane defined by basis points A, B, and C, as in Figure 6.15. To get its x-coordinate, we follow Equation 6.21

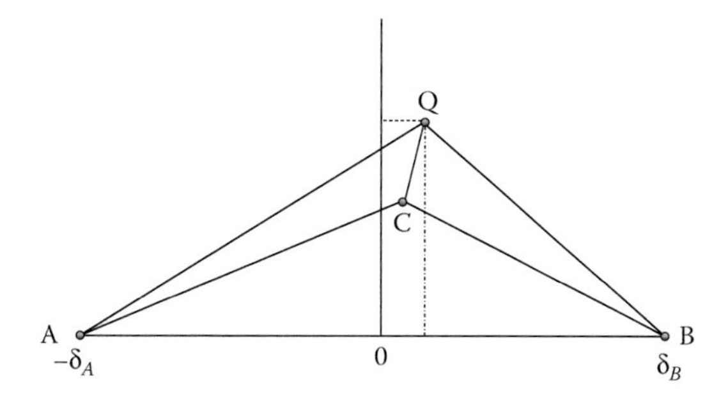

FIGURE 6.15 Consider an arbitrary point Q, whose distance to the basis points A, B, and C are given. The coordinates of point Q will be computed following the scheme described earlier.

$$x_Q = \frac{\delta_{AQ}^2 - \delta_{BQ}^2}{2\delta_{AB}} \tag{6.24}$$

Given the coordinates of points C and Q, for distance we get

$$\delta_{CQ}^2 = (x_C - x_Q)^2 + (y_C - y_Q)^2$$

which upon simplification may be expressed as

$$2y_C y_Q = x_C^2 + y_C^2 + x_Q^2 + y_Q^2 - \delta_{CQ}^2 - 2x_C x_Q$$

Using Equations 6.23 and 6.22c, we obtain

$$2y_C y_Q = \frac{\delta_{AC}^2}{2} - \frac{\delta_{AB}^2}{2} + \frac{\delta_{BC}^2}{2} + \frac{\delta_{AQ}^2}{2} + \frac{\delta_{BQ}^2}{2} - \delta_{CQ}^2 - 2x_C x_Q$$

or

$$y_Q = \frac{\delta_{AC}^2 - \delta_{AB}^2 + \delta_{BC}^2 + \delta_{AQ}^2 + \delta_{BQ}^2 - 2\delta_{CQ}^2 - 4x_C x_Q}{4y_C} \tag{6.25}$$

We have thus obtained the x- and y-coordinates of Q, given its distance to three basis points A, B, and C.

6.2.5 3D Space

Finally, let us consider 3D space. We relabel point Q as D and place it in 3D space, keeping its x- and y-coordinates unchanged.

Alternatively, Q may be any point, but its distances to A, B, and C are known. The basis points (four) of 3D space are A, B, C, and D. We can show this as in Equation 6.21

$$x_D = \frac{\delta_{AD}^2 - \delta_{BD}^2}{2\delta_{AB}} \tag{6.26a}$$

$$y_D = \frac{\delta_{AC}^2 - \delta_{AB}^2 + \delta_{BC}^2 + \delta_{AD}^2 + \delta_{BD}^2 - 2\delta_{CD}^2 - 4x_C x_D}{4y_C} \quad (6.26b)$$

$$z_D = \sqrt{\frac{\delta_{AD}^2}{2} - \frac{\delta_{AB}^2}{4} + \frac{\delta_{BD}^2}{2} - x_D^2 - y_D^2} \quad (6.26c)$$

Let Q be any point in 3D space with known distances to four basis points (A, B, C, and D). We will show that the coordinates of Q can be determined from available information. From Equations 6.26a and 6.26b, we determine the x- and y-coordinates

$$x_Q = \frac{\delta_{AQ}^2 - \delta_{BQ}^2}{2\delta_{AB}} \quad (6.27a)$$

$$y_Q = \frac{\delta_{AC}^2 - \delta_{AB}^2 + \delta_{BC}^2 + \delta_{AQ}^2 + \delta_{BQ}^2 - 2\delta_{CQ}^2 - 4x_C x_Q}{4y_C} \quad (6.27b)$$

To obtain the z-coordinate, we make use of the distance from Q to D

$$\delta_{QD}^2 = (x_Q - x_D)^2 + (y_Q - y_D)^2 + (z_Q - z_D)^2$$

$$z_Q = \frac{x_D^2 + y_D^2 + z_D^2 + x_Q^2 + y_Q^2 + z_Q^2 - \delta_{DQ}^2 - 2x_D x_Q - 2y_D y_Q}{2z_D} \quad (6.28)$$

Note that from Equation 6.22c we have the following result:

$$x_D^2 + y_D^2 + z_D^2 = \frac{\delta_{AD}^2}{2} - \frac{\delta_{AB}^2}{4} + \frac{\delta_{BD}^2}{2} \quad (6.29a)$$

Similarly, by replacing D with Q we have a symmetric result

$$x_Q^2 + y_Q^2 + z_Q^2 = \frac{\delta_{AQ}^2}{2} - \frac{\delta_{AB}^2}{4} + \frac{\delta_{BQ}^2}{2} \quad (6.29b)$$

Using Equations 6.29a and 6.29b in Equation 6.28, we get

$$z_Q = \frac{\delta_{AD}^2 + \delta_{BD}^2 + \delta_{AQ}^2 + \delta_{BQ}^2 - \delta_{AB}^2 - 2\delta_{DQ}^2 - 4x_D x_Q - 4y_D y_Q}{4z_D} \quad (6.30)$$

Thus, we can obtain the coordinates of an arbitrary point based on the knowledge of its distances to the basis points, whose coordinates are already known.

EXAMPLE 6.4

As an illustration of the previously mentioned method of finding the coordinates of an arbitrary point, given its distance to four basis points (3D space): Let the basis points be

$$A = [-5, 0, 0]$$

$$B = [5,0,0]$$

$$C = [6,8,0]$$

$$D = [7,8,5]$$

Consider a point Q whose distance to the previous basis points is

$$\delta_{AQ} = 296$$

$$\delta_{BQ} = 116$$

$$\delta_{CQ} = 77$$

$$\delta_{DQ} = 17$$

where δ_{AQ} stands for distance from point Q to the basis point A and so on. By solving Equations 6.27a, 6.27b, and Equation 6.30, we obtain the coordinates of Q as (9, 6, 8). These are indeed the coordinates assumed for point Q used for computing the distances to basis points. We have similarly obtained the correct results for many other arbitrary points in 3D space.

6.2.6 Missing Data

In any large sensor array, there will be some sensor pairs for which no reliable ToA measurements are available. The entries in the B matrix will be highly erroneous. Such entries may be replaced with estimates obtained from the basis-point method. As an example, consider the network shown in Figure 6.16.

Now, if we can estimate the coordinates of sensor 5 using the basis-point method, the distances to all unconnected sensors can be computed. For example, let sensor 5 be unconnected from some sensors in the array. Either the ToA measurements are not available or they are highly inaccurate. This situation may be corrected by choosing a set of basis points securely connected with sensor 5. We need three basis sensors, which are connected with sensor 5, with a reliable ToA. In Figure 6.16, we assume that sensors 1, 3, and 4 are connected with sensor 5. We select sensors 1, 3, and 4 as the basis points. We compute the x- and y-coordinates of sensor 5 using Equations 6.24 and 6.25. This procedure is repeated at all sensors, which happen to be only partially connected.

Next, using the computed coordinates, we compute distances between the partially connected sensors and complete the distance matrix required for executing the MDS algorithm.

6.2.7 Computational Load

In a standard MDS algorithm, the **B** matrix has M*(M−1)/2 inter-sensor distances to be computed from TOA measurements. In the present method, for p-dimensional space we need to compute (p+1) distances between all the basis points and of (p+1)

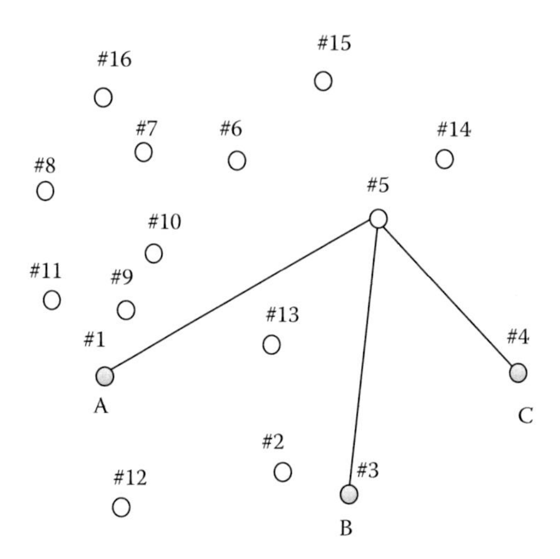

FIGURE 6.16 Consider a 16-sensor distributed array where a few sensors are not connected with some sensors, e.g., sensor 5 is one such sensor not connected with 13, 10, and 9. Therefore, the corresponding entries in the B matrix will have to be set to zero.

(M−p−1) distances between each basis point and (M−p−1) remaining sensors. Thus, the number of measurements in the basis-point method is linear with the number of sensors, but it is quadratic in the standard MDS method. The coordinates derived from the basis-point approach are only preliminary and they need to be refined by applying the standard MDS algorithm [11]. The coordinate estimates obtained through the MDS algorithm are superior as we use all available data instead of just four points in the basis-point method, where there may be significant errors in TOA estimation. The error in the basis-point method is reduced through a proper choice of basis points. Actual numerical experiments have shown that the error is reduced when the basis points surround most of the sensors [11,12].

In the presence of noise or errors in the estimation of distance, the rank of the B matrix will be equal to M, that is, full rank, although the first three eigenvalues will be dominating and the corresponding eigenvectors will yield optimum estimates in the sense that

$$\sum_{i=1}^{M}\sum_{j=1}^{M}(D_{i,j}^2 - \hat{D}_{i,j}^2)^2 = \min$$

where $\hat{D}_{i,j}^2$ is the estimated distance [12].

6.3 CLOCK SYNCHRONIZATION

Another significant limitation in ToA estimation arises from imprecise clock synchronization. Reported precision is on the order of 10 μs [13], which is adequate for acoustic signals but not for RF. Time-based ranging requires clock accuracy better than 1 ns, when 1 cm accuracy is desired. The local time of a clock in a sensor can be expressed as function $\gamma(t)$ of the true time t, where $\gamma(t) = t$ for a perfect clock. Over a short time interval, $\gamma(t)$ can be modeled as [14]

$$\gamma(t) = (1+\delta)t + \mu \tag{6.31}$$

where δ is clock drift, $d\gamma(t)/dt = 1+\delta$, and μ is clock offset (several μs). Clock drift affects the time interval measurement. If a single sensor wants to measure a time interval $\tau = t_2 - t_1$ sec, then the estimated time interval is

$$\hat{\tau} = \gamma(t_2) - \gamma(t_1) = (1+\delta)\tau \tag{6.32}$$

On the other hand, if a sensor wants to generate a delay τ_d, it effectively generates a delay of [15]

$$\hat{\tau}_d = \frac{\tau_d}{(1+\delta)} \tag{6.33}$$

6.3.1 Single Reply

Now let us consider two sensors communicating with each other. A packet of data is transmitted by sensor A at its local time $\gamma_A(t_1)$ and received by sensor B at its local time $\gamma_B(t_2)$. Sensor B calculates the estimated propagation delay as

$$\hat{\tau}_0 = \gamma_B(t_2) - \gamma_A(t_1)$$

$$= \tau_0 + \delta_B t_2 - \delta_A t_1 + \mu_B - \mu_A$$

where τ_0 $(= t_2 - t_1)$ is the true propagation time, including send time, access time, propagation time, and receive time, as illustrated in Figure 6.17, and $\hat{\tau}_0$ is the estimated propagation time. Sensor B retransmits the packet after a delay of τ_d but with an effective delay of $\hat{\tau}_d = \tau_d/(1+\delta_B)$.

The ToA at sensor A according to its clock is

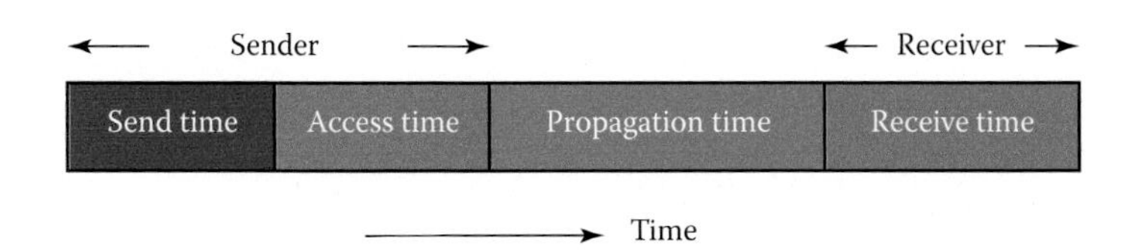

FIGURE 6.17 Breakdown of packet delay. $\tau_0 =$ send time + access time + propagation time + received time. (From Roche, M., *Time Synchronization in Wireless Networks*, 2006. Available online at: http://www1.cse .wustl.edu/~jain [16].)

$$\hat{\tau}_{RT1} = (2\tau_0 + \tau_d/(1+\delta_B))(1+\delta_A)$$
$$= 2\tau_0(1+\delta_A) + \tau_d \frac{(1+\delta_A)}{(1+\delta_B)} \tag{6.34}$$

where $\hat{\tau}_{RT1}$ stands for estimated return time. In the absence of any other information, sensor A will estimate the propagation time $\hat{\tau}_0$ as follows:

$$2\hat{\tau}_0 = 2\tau_0(1+\delta_A) + \tau_d \frac{(1+\delta_A)}{(1+\delta_B)} \tag{6.35}$$

The error in the estimate of propagation time is

$$\hat{\tau}_0 - \tau_0 = \tau_0\delta_A + 0.5\frac{(\delta_A - \delta_B)}{(1+\delta_B)}\tau_d \tag{6.36}$$

Clearly, the error is largely dominated by the response delay, which is on the order of several μs, whereas the propagation delay is on the order of a few tens of ns.

6.3.2 Double Reply

The second term in Equation 6.36 can be deleted through a simple step: Sensor B retransmits the packet after a delay equal to the response delay, $\tau_d/(1+\delta_B)$, according to its clock. This is called double reply, as shown in Figure 6.18 [17]. The return time of the second transmission is given by (Figure 6.19)

$$\hat{\tau}_{RT2} = (2\tau_0 + \tau_d/(1+\delta_B))(1+\delta_A) + \tau_d \frac{(1+\delta_A)}{(1+\delta_B)} \tag{6.37}$$

From Equations 6.35 and 6.37, it is possible to eliminate the terms involving the response delay. We obtain

$$\hat{\tau}_0 = \frac{2\hat{\tau}_{RT1} - \hat{\tau}_{RT2}}{2} = \tau_0 + \delta_A\tau_0 \tag{6.38}$$

If we assume that δ_A is known at sensor A, we can estimate the delay response without any error (in a noise-free case) as $\hat{\tau}_0/(1+\delta_A)$.

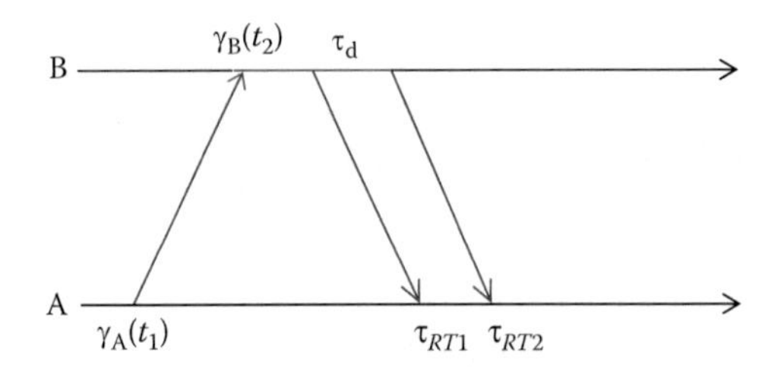

FIGURE 6.18 Single reply and double reply. Sensor B retransmits with a delay $\tau_d/(1+\delta_B)$.

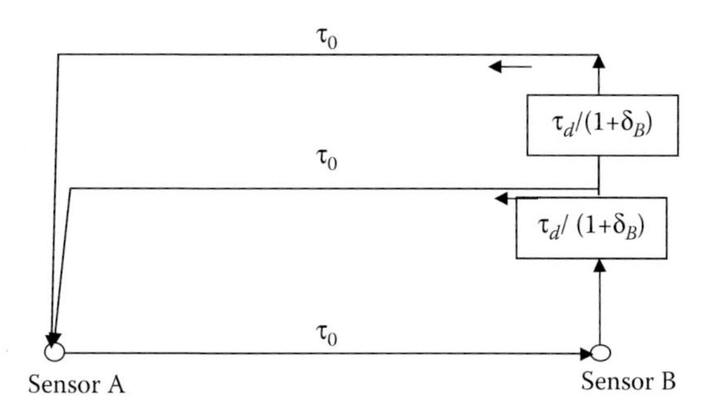

FIGURE 6.19 Double-reply scheme. The sensor B retransmits the received signal twice with a delay τ_d (effectively $\tau_d/(1+\delta_B)$) each time. τ_0 is the propagation delay.

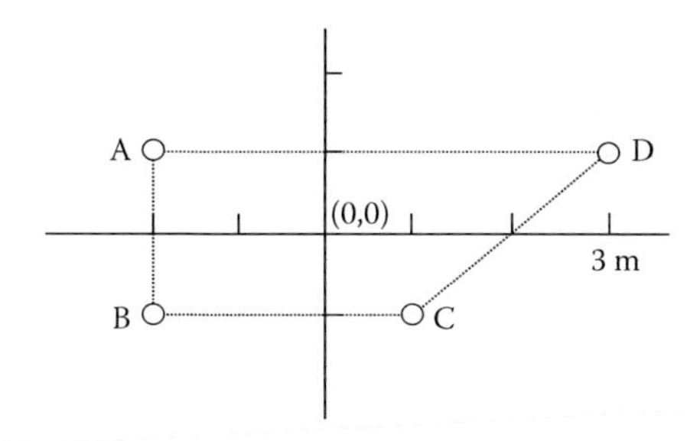

FIGURE 6.20 Four-sensor array.

EXAMPLE 6.5

This example demonstrates dramatic improvement in ToA/range estimation after using a double-reply scheme. The assumed parameters are

$$\delta_A = 0.001, \quad \delta_B = 0.002$$

$$c = 3\times 10^8 \text{ m/s}$$

and $\tau_d = 10^{-8}$ sec (10 ns). An array of four sensors are set apart, as in Figure 6.20. The relative distance between different sensor pairs is shown in a proximity matrix in Figure 6.21a. With single reply, the estimated range/ToA is shown in Figure 6.21b. There is a large error on account of the large response delay time (10 ns) in relation to propagation time. The situation has considerably been improved with the second reply (Figure 6.21c). In fact, an error-free range/ToA estimate is possible if the drift rate (δ_A) is known [18].

6.3.3 Linear Model

In general, the hardware clock of node A is a monotonically non-decreasing function of t. In practice, a quartz oscillator is used to generate the real-time clock. The oscillator's frequency depends on the ambient conditions, but for relatively extended

(a)

A	B	C	D	
0	2.0	3.6056	5.0	A
	0	3.0	5.3852	B
		0	2.6284	C
			0	D

(b)

0	4.9998	6.6062	8.0020
	0	6.0	8.3876
		0	5.6280
			0

(c)

0	2.002	3.6092	5.005
	0	3.003	5.3906
		0	2.6310
			0

FIGURE 6.21 Proximity matrices: (a) actual relative distances, (b) estimated relative distances after "single reply", (c) estimated relative distances after "double reply". Sensor response delay is 10 ns. All figures are in meters.

periods of time (minutes to hours), we can approximate with good accuracy using an oscillator with a fixed frequency, as in Equation 6.31. For another node B, we have a similar linear equation but with different coefficients. We can relate the outputs of two nodes:

$$\gamma_A(t) = (1+\delta_A)t + \mu_A \tag{6.39}$$

$$\gamma_B(t) = (1+\delta_B)t + \mu_B$$

Eliminating variable t from the previous equations

$$\gamma_B(t) = \frac{(1+\delta_B)}{(1+\delta_A)}\gamma_A(t) + \left(\mu_B - \frac{(1+\delta_B)}{(1+\delta_A)}\mu_A\right) \tag{6.40}$$

where

$$\frac{(1+\delta_B)}{(1+\delta_A)} > 0: \text{ Relative drift}$$

and

$$\left(\mu_B - \frac{(1+\delta_B)}{(1+\delta_A)}\mu_A\right): \text{ Relative offset}$$

For perfectly synchronized nodes $\delta_A = \delta_B$ and $\mu_A = \mu_B$, then $\gamma_A(t) = \gamma_B(t)$. In general, this situation is unlikely; we will have to match the time output of an arbitrary node with that of a reference node. Node B is synchronized with node A when the clock in B is matched to that in A. We can do this by estimating the relative drift and relative offset of B at different time instants. The minimum time instants required are two, but in the presence of measurement errors, we shall need many more measurements and employ the least-squares approach. At the stage of calibration, before deploying sensors to form an array, synchronization can be performed using the previous method. But once the sensors are spread out to their respective positions (not known), there is an extra factor of unknown propagation time. Then, the previous method will not work. Another method for synchronization is proposed in [19,20], where the sensors are spread apart.

Here we give an example of how Noh's approach works for two nodes, where one (node A) has an ideal clock (no drift ($\delta = 0$) and no offset ($\mu = 0$)). The second node (node B) has a normal clock (with drift and offset), and then our aim is to determine its unknown clock parameters (Figure 6.22).

In Figure 6.22,

$$U = \tau_0 + x + \mu_B$$

and

$$V = \tau_0 + y - \mu_B$$

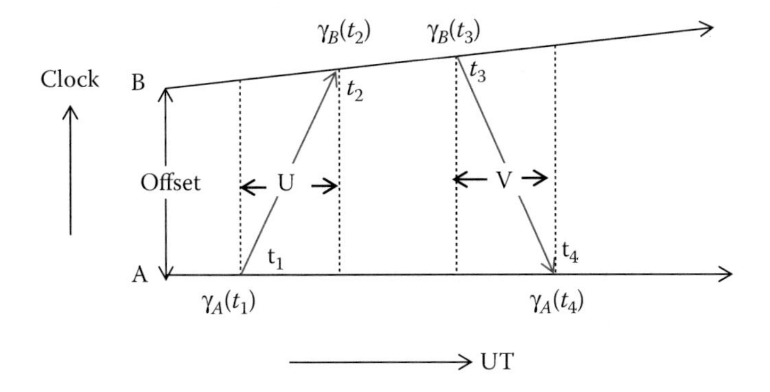

FIGURE 6.22 For ease of representation, we assume that node A has an ideal clock (no drift and no offset), but node B has both drift and offset. UT: universal time (x-axis) and Clock: clock output (y-axis).

Recall that τ_0 stands for total delay response time (send time + access time + propagation time + receive time, see Figure 6.17) and x and y stand for random delays. The time intervals, in terms of the observed time stamps imprinted on the test packet of data as it leaves the A node and reaches the B node and again as it leaves the B node and returns to the A node, are given by

$$U = \gamma_B(t_2) - \gamma_A(t_1) = \tau_0 + x + \mu_B \tag{6.41}$$

$$V = \gamma_B(t_3) - \gamma_A(t_4) = \tau_0 + y - \mu_B$$

Assuming no drift, offset may be estimated from Equation 6.41 after averaging over many observations,

$$\hat{\mu}_B = \frac{\bar{U} - \bar{V}}{2} + \frac{\bar{x} - \bar{y}}{2} \tag{6.42}$$

and

$$\hat{\tau}_0 = \frac{\bar{U} + \bar{V}}{2} + \frac{\bar{x} + \bar{y}}{2}$$

where overhead bars stand for the average taken over a large number of measurements. Thus, both delay and offset can be estimated from $\bar{U}$ and $\bar{V}$ [21].

6.3.4 Joint Estimation of Offset, Skew (Drift), and Delay

Next, we shall bring in the drift factor:

$$\bar{U} = \bar{\gamma}_B(t_2) - \bar{\gamma}_A(t_1) = \delta_B(\bar{\gamma}_A(t_1) + \tau_0 + \bar{x}) + \tau_0 + \bar{x} + \mu_B \tag{6.43}$$

$$\bar{V} = \bar{\gamma}_B(t_3) - \bar{\gamma}_A(t_4) = \delta_B(\bar{\gamma}_A(t_4) - \tau_0 + \bar{y}) - \tau_0 + \bar{y} + \mu_B$$

Adding, we get

$$\hat{\mu}_B = \frac{\bar{U} + \bar{V}}{2} - \delta_B\left(\frac{\bar{\gamma}_A(t_4) + \bar{\gamma}_A(t_1)}{2}\right) - (1 + \delta_B)\frac{\bar{x} + \bar{y}}{2} \tag{6.44a}$$

and subtracting. we get

$$\frac{\bar{U} - \bar{V}}{2} = \delta_B\left(\frac{\bar{\gamma}_A(t_1) - \bar{\gamma}_A(t_4)}{2} + \tau_0\right) + \tau_0 + (1 + \delta_B)\frac{\bar{x} - \bar{y}}{2} \tag{6.44b}$$

Solving Equations 6.44a and b under the assumption, appropriate for large number of observations, $\bar{x} = \bar{y} = 0$ (or in a noise-free case), we can easily obtain δ_B and μ_B as follows:

$$\delta_B = \frac{\dfrac{\bar{U} - \bar{V}}{2} - \tau_0}{\left(\dfrac{\bar{\gamma}_A(t_1) - \bar{\gamma}_A(t_4)}{2} + \tau_0\right)} \tag{6.45a}$$

and

$$\mu_B = \frac{\bar{U}+\bar{V}}{2} - \left(\frac{\bar{\gamma}_A(t_4)+\bar{\gamma}_A(t_1)}{2}\right)\frac{\frac{\bar{U}-\bar{V}}{2}-\tau_0}{\left(\frac{\bar{\gamma}_A(t_1)-\bar{\gamma}_A(t_4)}{2}+\tau_0\right)} \tag{6.45b}$$

Expressing Equation 6.43 in a matrix form, taking the average of all measurements we have

$$\begin{bmatrix} \gamma_B(t_2) \\ \\ \gamma_B(t_3) \end{bmatrix} = \begin{bmatrix} 1 & (\bar{\gamma}_A(t_1)+\tau_0+\bar{x}) \\ \\ 1 & (\bar{\gamma}_A(t_4)-\tau_0+\bar{y}) \end{bmatrix} \begin{bmatrix} \mu_B \\ \\ (1+\delta_B) \end{bmatrix} \tag{6.46a}$$

$$\begin{bmatrix} \gamma_B(t_2) \\ \\ \gamma_B(t_3) \end{bmatrix} = \mathbf{H} \begin{bmatrix} \mu_B \\ \\ (1+\delta_B) \end{bmatrix} \tag{6.46b}$$

where

$$\mathbf{H} = \begin{bmatrix} 1 & (\bar{\gamma}_A(t_1)+\tau_0+\bar{x}) \\ 1 & (\bar{\gamma}_A(t_4)-\tau_0+\bar{y}) \end{bmatrix}$$

We can now easily solve Equation 6.46 for offset and drift

$$\begin{bmatrix} \mu_B \\ (1+\delta_B) \end{bmatrix} = \mathbf{H}^{-1} \begin{bmatrix} \bar{\gamma}_B(t_2) \\ \bar{\gamma}_B(t_3) \end{bmatrix} \tag{6.47a}$$

where the inverse of the **H** matrix is easily obtained as

$$\mathbf{H}^{-1} = \frac{\begin{bmatrix} (\bar{\gamma}_A(t_4)-\tau_0+\bar{y}) & -(\bar{\gamma}_A(t_1)+\tau_0+\bar{x}) \\ \\ -1 & 1 \end{bmatrix}}{\det(\mathbf{H})} \tag{6.47b}$$

and its determinant is given by

$$\det(\mathbf{H}) = -2\left((\bar{\gamma}_A(t_1)-\bar{\gamma}_A(t_4))/2+\tau_0-\frac{\bar{x}-\bar{y}}{2}\right)$$

Using the expression for the determinant in Equation 6.47b and then the inverse expression in Equation 4.47a, we can solve for the unknown parameters. Note that the actual delay time τ_0 is known beforehand:

$$1+\delta_B = \frac{(\bar{\gamma}_B(t_2)-\bar{\gamma}_B(t_3))/2}{\left((\bar{\gamma}_A(t_1)-\bar{\gamma}_A(t_4))/2+\tau_0-\frac{\bar{x}-\bar{y}}{2}\right)} \tag{6.48a}$$

$$\mu_B = \frac{\begin{bmatrix} \overline{\gamma}_B(t_3)\overline{\gamma}_A(t_1) - \overline{\gamma}_A(t_4)\overline{\gamma}_B(t_2) + (\overline{\gamma}_B(t_2) + \overline{\gamma}_B(t_3))\tau_0 \\ +\overline{x}\overline{\gamma}_B(t_3) - \overline{y}\overline{\gamma}_B(t_2) \end{bmatrix}}{2\left((\overline{\gamma}_A(t_1) - \overline{\gamma}_A(t_4))/2 + \tau_0 - \dfrac{\overline{x} - \overline{y}}{2} \right)} \tag{6.48b}$$

where we have assumed that measurement errors are uncorrelated with clock outputs. Solutions of Equation 6.48 reduce to Equation 6.45, which was derived assuming $\overline{x} = \overline{y} = 0$.

EXAMPLE 6.6

Refer to Figure 6.22. Node A is an ideal sensor with zero offset (μ_A) and zero skew $\left(\hat{\delta}_A\right)$. Node B has both offset, μ_B=0.001 sec, and skew, $\hat{\delta}_B = 0.11/\text{sec}$. Packet transmission is time $t_1 = 0.0$, which is also reference time, and packet arrival time is $t_4 = 1.5$ sec. In the first instance, we assumed a no-noise condition. Estimates of U and V, which are shown in Figure 6.22, are respectively, U = 0.0010211 and V = 1.6660. The solution of Equation 6.47 yields $\mu_B = 0.0010$ and Δ_B = 1.1100, as expected.

Next, we introduce random errors, x and y, in the delay estimation in U and V. Both errors are zero mean and variance 10^{-8} Gaussian random variables. We have computed U and V for different numbers of observations and then have estimated $\hat{\delta}_B$ and $\left(\hat{\delta}_A\right)$. The results are shown in Table 6.3. They are only representative results averaged over 100 experiments. No attempt is made to derive statics in terms of their mean and variance. Noh [19] has already provided extensive theoretical derivations of mean and variance and simulation results.

It may be seen that the drift estimation is more stable, even with as few as ten observations. We have assumed that the delay is known.

But this is not so when the range of an unknown node is concerned.

6.3.5 Estimation of Travel Time

The time of travel is not known. It is, however, important that we estimate this quantity and thereby get an estimate of the range. An easy approach is to compute $\hat{\mu}_B$ and $\hat{\delta}_B$ from Equation 6.45 for a series of delays covering the expected range. It is found that $\hat{\mu}_B$ remains constant but $\hat{\delta}_B$ decreases with increasing delay, crossing the

TABLE 6.3
Actual Values of μ_B and δ_B Are 0.001 s and 1.11/s, Respectively

N →	10	100	1000
$\hat{\delta}_B$ →	1.1078	1.1090	1.1103
$\hat{\mu}_B$ →	0.001088	0.00321	0.00126

Note: N stands for number of observations used for averaging U and V.

zero line at the correct delay. We demonstrate this mathematically in Equation 6.45 and through simulation.

First, we show that μ_B is always a constant for any $\tau = \tau_0 + \Delta$, where $|\Delta| \ll \tau_0$, by virtue of the assumed parameters, namely, $\mu_A = 0.0$ and $\delta_A = 0.0$. Equation 6.48b reduces to

$$\tilde{\mu}_B = \frac{\left[\varepsilon_1 + (\bar{\gamma}_B(t_2) + \bar{\gamma}_B(t_3))\tau_0 + (\bar{\gamma}_B(t_2) + \bar{\gamma}_B(t_3))\Delta\right]}{2(\varepsilon_2 + \tau_0 + \Delta)}$$

where

$$\varepsilon_1 = \gamma_B(t_3)\gamma_A(t_1) - \gamma_A(t_4)\gamma_B(t_2)$$

and

$$\varepsilon_2 = (\gamma_A(t_1) - \gamma_A(t_4))/2\,.$$

It is found that

$$\varepsilon_1 + (\bar{\gamma}_B(t_2) + \bar{\gamma}_B(t_3))\tau_0 \gg (\bar{\gamma}_B(t_2) + \bar{\gamma}_B(t_3))\Delta$$

and

$$\varepsilon_2 + \tau_0 >> \Delta\,;$$

hence, $\tilde{\mu}_B = \mu_B$ independent of Δ.

Next, we look into the variation of δ_B in the neighborhood of τ_0,

$$\tilde{\delta}_B = \frac{\dfrac{\bar{U} - \bar{V}}{2} - \tau_0 - \Delta}{\left(\dfrac{\bar{\gamma}_A(t_1) - \bar{\gamma}_A(t_4)}{2} + \tau_0 + \Delta\right)} \tag{6.49a}$$

where Δ is the difference between the assumed delay and the actual delay (unknown). Using Equation 6.44b in Equation 6.49a, we obtain

$$\tilde{\delta}_B = \frac{\delta_B\left(\dfrac{\bar{\gamma}_A(t_1) - \bar{\gamma}_A(t_4)}{2} + \tau_0\right) - \Delta}{\left(\dfrac{\bar{\gamma}_A(t_1) - \bar{\gamma}_A(t_4)}{2} + \tau_0 + \Delta\right)} = \frac{\delta_B - \Delta / L}{(1 + \Delta / L)} \tag{6.49b}$$

where

$$L = \left(\frac{\bar{\gamma}_A(t_1) - \bar{\gamma}_A(t_4)}{2} + \tau_0\right)$$

and $\Delta \ll |L|$. Using the first-order approximation, Equation 6.49b may be reduced to

$$\begin{aligned}\tilde{\delta}_B &\simeq \left(\delta_B - \Delta/|L|\right)\left(1-\Delta/|L|\right) \\ &= \delta_B - \Delta/|L| - \delta_B\Delta/|L| + \left(\Delta/|L|\right)^2\end{aligned} \tag{6.49c}$$

Dropping the second order terms in Equation 6.49c, we obtain a result valid around the actual delay

$$\tilde{\delta}_B - \delta_B \approx -\Delta/|L| \tag{6.50}$$

Equation 6.50 implies that

$$\tilde{\delta}_B - \delta_B > 0 \text{ for } \Delta < 0$$

and

$$\tilde{\delta}_B - \delta_B < 0 \text{ for } \Delta > 0$$

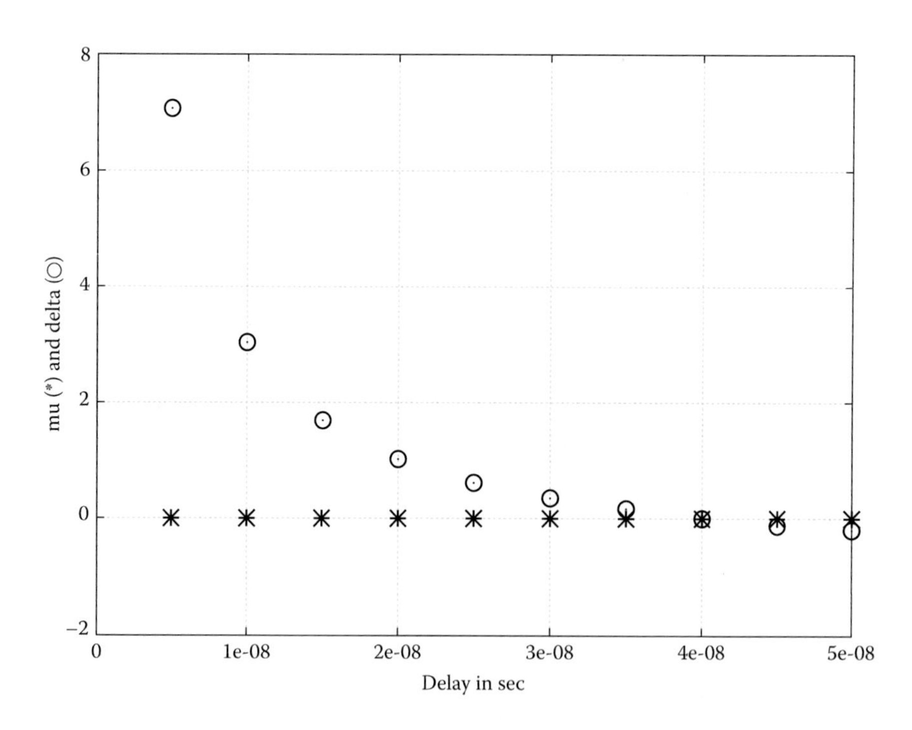

FIGURE 6.23 For the correct selection of delay time, estimates of delta and mu are exact. As delay is varied from 5 ns to 50 ns, the estimates of delta decrease from positive to negative values at about true delay, in this example at tau = 40 ns. Estimates of mu (*) remain unchanged throughout.

$\tilde{\delta}_B - \delta_B$ changes its sign at $\Delta = 0$. We have used this property to estimate τ_0, as demonstrated in Figure 6.23. The parameters used for the previous computation are as follows:

$$\mu_A = 0.0 \quad \delta_A = 0.0$$

$$\mu_B = 10 \text{ n sec}$$

$$\delta_B = 0.01/\text{sec}$$

$$\tau_0 = 40 \text{ n sec}$$

$t_1 = 0.0$ and $t_4 = 1.0$ (see Figure 6.22).

Observation errors are Gaussian random variables with zero mean and variance equal to 10^{-8}. The number of observations is ten. Estimated $\hat{\mu}_B$ and $\hat{\delta}_B$ were averaged over 100 independent experiments. We can thus jointly estimate offset, drift, and delay.

6.4 SUMMARY

In the first section, we considered self-localization without any help from an external transmitter. The key requirement is the presence of a sufficient number of anchor nodes, such that most sensors are within the reach of at least three anchors. Sensors thus localized act as anchors for other sensors. In this way, a large DSA may be localized even with a few blanks. In the next section, we proposed localization without any anchor nodes. We apply the well-known procedure of MDS from statistics to localization using distance information only. Unfortunately, the method leads to some uncertainties, such as translation, rotation, and mirror imaging of sensors, which cannot be rectified without anchor nodes or some other information about the network.

Time synchronization of all nodes and anchors is essential, considering that the clocks in nodes, being inexpensive, tend to have large offset and drift. An anchor node transmits a message to a node within its range. The message is time stamped by both nodes according to their own clocks. The simple difference of stamped time does not yield the correct delay. The message is returned to the same anchor node. The ToA still has a large error that can be eliminated only by retransmitting the same message after the same delay. The term for this is *double reply*. The error depends on the offset.

Next, we discuss joint estimation of the offset, drift (skew), and delay, after synchronizing the node nearest to the anchor node (whose clock apparently outputs universal time). In the presence of measurement errors, we need to repeat the experiment many times and average the node outputs. Clock frequency is assumed stable over the experiment duration. All three parameters, namely, offset, drift (skew), and delay times, are computed using the arrival time of data packets.

REFERENCES

1. J. N. Ash and R. L. Moses, On optimal anchor node placement in sensor localization by optimization of subspace principal angles, *ICASSP*, ICASSP_2008, pp. 2289–2292, 2008.
2. C. Savarese, J. M. Rabaey, and J. Beutel, Locationing in distributed ad-hoc wireless sensor networks, *IEEE ICASSP*, pp. 2037–2040, 2001.
3. P. S. Naidu, *Modern Digital Signal Processing: An Introduction*, 2nd ed., New Delhi: Norosa, 2006.
4. F. K. W. Chan, H. C. So, and W.-K Ma, A novel subspace approach for cooperative localization in wireless sensor networks using range measurements, *IEEE Transactions on Signal Processing*, vol. 57, no. 1, pp. 263–269, 2009.
5. W. S. Torgerson, Multidimensional scaling: I. theory and method, *Psychometrika*, vol. 17, pp. 401–419, 1952.
6. J. B. Kruskal, Multidimensional scaling by optimizing goodness of fit to a nonmetric hypothesis, *Psychometrika*, vol. 29, pp. 1–27, 1964.
7. T. F. Cox and M. A. A. Cox, *Multidimensional Scaling*, 2nd ed. Boca Raton, FL: Chapman & Hall, 2001.
8. X. Ji and H. Zha, Sensor positioning in wireless ad-hoc sensor networks using multidimensional scaling, *IEEE*, pp. 2652–2661, 2004.
9. S. T. Birchfield and A. Subramanya, Microphone array position calibration by basis-point classical multidimensional scaling, IEEE transactions on speech and audio processing, 2004.
10. F. W. Young and N. Cliff, Interactive scaling with individual subjects, *Psychometrika*, vol. 37, pp. 385–415, 1972.
11. A. Subramanya and S. T. Birchfield, Extension and evaluation of MDS for geometric microphone array calibration, European signal processing conference (EUSIPCO), Sept. 2004.
12. S. T. Birchfield, Geometric microphone array calibration by multidimensional scaling, *IEEE ICASSP*, ICASSP_2003, pp. V1657–160, 2003.
13. N. Patwari, J. N. Ash, S. Kyperountas, A. O. Hero III, R. L. Mose, and N. S. Correal, Locating the nodes, *IEEE Signal Processing Magazine*, pp. 57–69, July 2005.
14. F. Sivrikaya and B. Yener, Time synchronization in sensor networks: A survey, *IEEE Network*, pp. 45–50, 2004.
15. D. Dardari, U. Ferner, A. Giogetti, and M. Z. Win, Ranging with ultra wide bandwidth signals in multipath environments, *Proceedings of the IEEE*, vol. 97, no. 2, pp. 400–426, 2009.
16. M. Roche: Time synchronization in wireless networks, 2006. Available at: http://www1.cse .wustl.edu/~jain.
17. J. H. Hwang and J. M. Kim, Double reply ToA algorithm robust to timer offset for UWB system, *IEICE Transactions*, vol. 92B, pp. 2218–2221, 2009.
18. D. Venkataramana, Double reply TOA algorithm robust to timer offset for UWB system (Project report submitted for award of B. Tech degree), Department of Electronics and Communications Engineering, MVGR College of Engineering, Vizianagaram, Andhra Pradesh. India, 2009.
19. K.-L. Noh, Q. M. M. Chaudhari, E. Serpendin, and B. W. Suter, Novel clock phase offset and skew estimation using two-way timing message exchanges for wireless sensor networks, *IEEE Transactions on Communications*, vol. 55, pp. 766–777, 2007
20. K.-L. Noh, E. Serpendin, and K. Qaraqe, A new approach for time synchronization in wireless sensor networks: Pairwise broadcast synchronization, *IEEE Transactions on Wireless Communications*, vol. 7, pp. 3318–3322, 2008.
21. S. Ganeriwal, R. Kumar, and M. B. Srivastava, Timing: Synchronization protocol for sensor networks, *SenSys*, Nov. 5–7, Los Angeles, pp. 138–149, 2003.

Index

P

Q

R

S

T

U

V

W

Z